The Sasquatch:

Journey Through the Veil

By

Samantha Ritchie

COPYRIGHT PAGE
ISBN-13:978-1530762705

THE SASQUATCH: JOURNEY THROUGH THE VEIL
COPYRIGHT 2016 SAMANTHA RITCHIE
ALL RIGHTS RESERVED
INTERNATIONAL COPYRIGHT PROTECTION IS RESERVED UNDER UNIVERSAL COPYRIGHT CONVENTION AND BILATERAL COPYRIGHT RELATIONS OF THE USA. ALL RIGHTS RESERVED, WHICH INCLUDES THE RIGHT TO REPRODUCE THIS BOOK OR ANY PORTIONS THEREOF IN ANY FORM WHATSOEVER EXCEPT AS PROVIDED BY RELEVANT COPYRIGHT LAWS.

WWW.CREATESPACE.COM
PRINTED IN USA
ISBN-13:978-1530762705
PUBLISHED BY:
GREENWATER PUBLISHING AND DISTRIBUTING

FRONT COVER PHOTO BY: BARBARA SHUPE

Dedicated to my mother who has always been there for me through all the good as well as the dark times of my life!

Table of Contents

Forward		6
Chapter 1	Introduction	7
Chapter 2	Meeting Barb Shupe	17
Chapter 3	Initial Vocalizations and Sightings	19
Chapter 4	Giant Log Slammed in the Ground	24
Chapter 5	My Psychic Awakening	27
Chapter 6	The Cloaking Bigfoot	36
Chapter 7	The Light Portal and Orb	39
Chapter 8	The Local Clan Observes Me	42
Chapter 9	Direct Sasquatch Interaction	48
Chapter 10	Backyard Encounter	53
Chapter 11	A Night Walk to Remember	56
Chapter 12	Barb and Gabby June 2015 Campout	57
Chapter 13	The Blue Mountain Campout	63
Chapter 14	Connecting with the Sasquatch	78
Chapter 15	Apparition or Sasquatch Mind Trick	81
Chapter 16	A Personal Message from the Sasquatch	85
Chapter 17	Morphing Bigfoot Incident	88
Chapter 18	Sasquatch and the UFO Incident	93
Chapter 19	Two Beings Appear Behind Me	101

Chapter 20	The Habituation Method	106
Chapter 21	Photographic Evidence	131
Chapter 22	Communicating with the Sasquatch	139
Chapter 23	Stop, Be Quiet, Listen!	141
Chapter 24	Human's Artificial World	144
Chapter 25	Know Yourself – Question Everything!	153
Chapter 26	The Worldwide Agenda of Fear	156
Chapter 27	Humanity's Time for Decisive Action	161
Appendix I	The 2012 Event	164
Appendix II	Personal Empowerment Message	168
Appendix III	Online Resources	170
Appendix IV	Young Girl's Report of a Sighting	172

Forward

Samantha Ellen Ritchie has been living an adventurous life full of diverse beliefs and experiences. It is no surprise to me, whatsoever, that she is able to author a book such as this one. Regarding the Bigfoot phenomena, she's been able to blend her beliefs and experiences together in a masterful way to help the reader walk in her shoes and view the Forest People through her eyes. Her book takes the reader to a whole new level of understanding while leaving the possibilities wide open because there's so much more for all of us to learn. Now please buckle up and enjoy the ride.

Happy reading!!!

- Dr. Matthew A. Johnson

Chapter 1
Introduction

If someone told me even ten years ago that I would be writing my first book about the Sasquatch because of my personal experiences with them, I would have told them they were out of their minds. The first thing I would have spouted out would have been "Bigfoot is not real, it's a crazy hoax someone came up with to make a few bucks!"

My earliest recollection of Bigfoot would have been in the 1970's after the Patterson event took place in 1967. It was a different day back then and information was very limited. Information was only available by written pages (books, magazines and newspapers); radio and only three TV channels representing the major networks were available at the time. What little bit I was exposed to were the stories that this was a hoax, including how the foot print casts were purportedly fake. Without any personal experience or factual information, it was dismissed out of my mind for decades. However everything in life happens for a reason and when the time is right things eventually get revealed to us. Perhaps I should start from the beginning...

I was born in the mid 1950's to a middle class family back in the state of Kentucky near the Fort Knox military base. I can honestly say in looking back, we were raised very well and did not lack materially. Dad worked very hard in the amusement (jukebox/pinball) business putting in long stressful hours while Mom stayed home doing a great job taking care of us.

When I was seven years old we moved from an older neighborhood into a brand new subdivision. Our new home was a pink brick house with four bedrooms and a basement. Dad soon built in two extra bedrooms in the basement along with a bar lounge area completely with our own jukebox. A lot of great New Year's Eve parties took place in those days!

The best thing about this home, behind the fenced backyard, was a forest (I called the woods) that stretched back for at least a couple of miles before coming to a county road. My Dad built a wooden step bridge over the fence so that we could take walks back there. I immediately fell in love with those woods! Instead of wanting to do other activities like playing ball or watching TV after school, as soon as I got home I leaped over that fence and would stay out there by myself until it was nearly dark or suppertime, whichever came first. It was during this time that I started to appreciate all the wonders of nature. Being empathic even as a child, I started to sense a feeling emanating from the trees and plants. It's was a strange calmness and the feeling of joy unlike any I would experience while at school from the other kids or adults. I believe this was when I become more of a hermit and would limit myself to having just a couple of close friends. Groups of kids terrified me so the woods basically become my sanctuary away from the mixed up emotions coming from others.

Soon I felt like a caretaker of the forest and would rake the leaves to create trails through different parts of the woods. I even made little signs marking each trail and a particular spot with its own unique name. I can only remember some of these like Woodland Trail and Christmas Hill. As I got a little older I would camp out there with close friends. It was an awesome experience to hear all the different sounds coming from locusts, crickets and different birds!

Though the 1960's were a peaceful time in my life it would soon end and would put me on a hard road until recently. In late November of 1969 I woke up in the middle of the night with extreme pain in my left leg. I tried to move but it was agonizing and ended up screaming for my parents. They called an ambulance and rushed me to the hospital. As I was checked into my room, the doctors were perplexed about the reason for my pain. What started out as one day soon dragged on through December as I got worse with my leg swelling up to three times it's normal size. I started having fever and at one point was as high as 107 degrees. I was also experiencing hallucinations such as the sensation of being above my own body. A Catholic Priest was finally called in to give me my last rites as I would soon be dead if nothing changed.

Dr. Colbert was finally contacted in Lousville. He came down to our hospital, took one looked at my leg and told the doctors "You need to open the leg up and let it drain – it's full of pus!"

It seemed I had contracted a rare disease called osteomylitis which causes infections in the bone marrow. I was later moved to the hospital in Loisville and stayed for the month of January until I was able to return home. Dr. Colbert saved my life and I am so grateful he came on the scene before it was too late!

After that experience something changed about me. All my straight, coarse hair fell out and was replaced by wavy hair. My personality was also in flux but then again, I was going into my teen years. Also, while I continued to go out in the woods, something was talking to me by using subtle mental thoughts causing me to deeply search for answers... the kind that schools or religion were not providing. Finally, life got in the way. My parents divorced and, for the rest of the 1970's, we had to move to different homes and away from the place I called my sanctuary.

This strong urge for soul searching caused me to look outside my Catholic faith for answers in different Christian denominations. For a brief time, I became baptised as a Baptist until a school friend of mine had a brother that was part of this unusual Christain religion radically different from the others. Instead of listening to my inner voice, I got caught up into the wonderful promises of a paradise for mankind. I would not have to think for myself to figure things out because now there was a religion that provided all the answers. Slowly, due to the constant drumming of their teachings in the three meetings a week we had to attend, my mind slowly closed and would not accept anything as reality unless it was in total agreement with the teachings of this religion.

So for the next thirty-four years I was mentally locked into this religious group. I ended up marrying a congregation member (who later left the faith) in Washington state and we had three children together. However after ten years together, we divorced and I remarried another member in North Carolina who helped raise my two younger children. Life during this time was anything but peaceful and it was everything I could do to keep things together mentally and emotionally.

I can't say that I was considered a "strong" member of that

faith as I didn't really agree with some of the things that were going on within these congregations but I didn't dare speak up and risk being excommunicated.

I did move back to Kentucky for a short five years from 1997 to 2002. At first it was blissful and I *thought* I would recapture some of that calmness I felt years ago as a child. I even bought a downtown building in the historic district of town, remodeled it myself and opened the Cobblestone's Cafe. During this time I was also handling my main job working as a software programmer for a major internet retailer. Add to that the pressure from the local congregation to conform to certain standards and my daily arguments with my spouse, I was just about at the end of my rope with no way out! Tragedy struck in the summer of 2001 when a well-known and well-liked local woman and her daughter died suddenly in Florida from a parasailing accident. She used to help me out with getting supplies for my place and now she was gone. Then 9-11 happened! Since we were in a town right next to Fort Knox the base was suddenly locked down and everyone was fearful to leave their homes. Hardly anyone was coming to the restaurant for at least two months. Something had to change in my life!

I immediately rented out the restaurant to a new business owner and started to lay low for a few months. Around September of 2002 I came into quite a bit of money from cashing out some stocks I had in the company I worked with. Trying to determine what to invest my money into, I attended a real estate auction and *almost* bought a house for cash. I was outdone by a persistent bidder and today I am so thankful she was there as it would have altered the course of events that would follow!

After that auction I woke up on a Sunday morning and something within me started speaking to me "You need to go back to Washington state". I told my spouse Toby that I was buying a plane ticket to visit Washington again and possibly look for a home to purchase." Toby was all for it so I left a couple of days later. Once I arrived in Seatac airport and checked into the hotel, I took a road trip around the Cascades. As I made my way through the mountains I felt that calm and euphoric feeling come over that I haven't felt since I was a kid. Though I was there only ten days I was able to locate a home in Tacoma and moved the family there six weeks later.

Once in Tacoma, I had the ability to telecommute for my job and would go to various coffee shops around town to work. I was still stuck in the city most of the time but every so often I would head out towards Mount Rainier to visit and hike. However, things again started going downhill when my daughters left to live with my first spouse. I missed them dearly and felt an empty space in my heart. Also, we were still connected to the religious group and that, no doubt, had a bearing on my daughters leaving.

By the middle of the first decade of the 2000's I was feeling very empty so I tried to become more engaged with the company I had been with for over sixteen years. I became a Vice-President of Research and Development and ended up attending large conventions, meeting quite a few celebrities and occasionally speaking on the stage to over twenty thousand people.

This of course did not improve my marriage and the arguments continued on a daily routine. Soon, it wouldn't be long before my life would change radically once again!

In May of 2007, while I was in North Carolina on company business, I had what some may call an emotional breakdown but it was more of a realization about who I was and something not being right about myself physically. It involved what is called "gender dysphoria" and that in itself would require a separate book concerning what I had to go through for the last nine years. When I arrived back in Tacoma I had to tell Toby what was up with me and that created some stressful situations. Then, while I was back in North Carolina attending a gala event in August, I got a call that my father just passed away in Kentucky. I left immediately and drove a rental car back there to attend the funeral. In September, when things couldn't get much worse, Toby was diagnosed with lung cancer and was given only three years to live.

The next three years were sheer hell on earth as we tried to find alternative treatments for the cancer. We found a clinic in Mexico and ended up briefly owning a home in Del Rio, TX where Toby lived while I was in Washington state. The treatments did not help so we sold that home and I moved Toby to North Carolina for conventional treatments of chemo and radiation and to be near family. I can say that we got along better during this time period than all the seventeen prior years we were married.

It was also a time period where I started to realize the religious belief system I had adhered to for the past 34 years as being the absolute truth was not the truth after all – not by a long shot! It was like waking up from living a lie your entire life. How could I have let this happen to me and for so many years? I was so sure that I knew the truth about God and the universe. I now realized that everything (and I do mean everything) I was ever taught was suspect and subject to be questioned. My closed mind that had stayed locked away in a very small box was now free! I now had an open mind about everything! This helped me to better deal with the challenges ahead.

Insurance now became the critical issue and in 2010 I had to move Toby back to Washington away from the family in North Carolina. Being on hospice from June to August, I watched as Toby withered away until finally passing away on September 5th, 2010. I was devastated! It was a tough go for the next couple of years but it also became the most enlightening time for me.

While dealing with the grief, I had to get away from the city and escape! I looked towards the Cascade Mountains for peace and was drawn, once again, to its energy. Since I was able to telecommute, I began to drive thirty miles to Enumclaw a couple of times a week for a few hours while I typed away on my laptop in the local Starbucks. On one of these visits, I was curious about what lay ahead towards the mountains on Hwy 410. I went ahead and drove the seventeen mile trip, coming to this place called Greenwater.

I was in awe of the scenery, with the White River running along the highway and the pleasant little shops within the village. At first, I would only visit the Federation Forest State Park and afterwards stop in at the Wapiti Woolies shop to browse around and get a cup of coffee. It was a very tranquil experience unlike anything I had in Tacoma. In the beginning, I would go drive there from Tacoma once a week then it became twice a week. I would leave earlier in the morning so that I could not only spend time in the forest but also be able to work for a number of hours hanging out at the local tavern while using their Wifi.

Visiting Greenwater was soon an obsession with me. It was as if something there was calling me to come and stay. I could also *feel* the energy coming from the river and the rocks I touched as if recharging my own spirit! These visits continued until an opportunity presented itself that I could not pass up! In mid July of 2012, an RV was parked in front of the tavern and was for sale at an extremely reasonable price. Without having to think about it, I immediately contacted the owner and bought it. Also a lot in the back of the parking lot of the tavern was available so I arranged to park my RV there and rent the spot on a month to month basis. Though I still had the home in Tacoma, I now was able to spend several days a week living in what I would call one of the most beautiful spots in the country!

I will mention that, for the past couple of years since Toby died right up through the end of 2012, I was going through a spiritual transformation of sorts. I had become very depressed about Toby's passing and was feeling deep loneliness to the point of my chest hurting. Also work was very stressful due to the responsibility I had as a corporate VP. I really needed answers to life's biggest questions – Who am I? Why am I here?

After leaving the religion for which I was mentally trapped for over thirty four years, I was in no mood to be controlled by any religious group. I didn't blame the Bible as much as I blamed the way these groups interpreted it to suit their own agendas.

Still being interested in studying the Bible, I found a small household that was doing that – so I thought. Turns out that they were part of another strange cult that felt the writings of this one woman were more important than the actual study of the Bible itself.

Not giving up on the Bible, I did allow myself to be open to other possibilities of connecting with God outside of religious organizations. As the year 2012 was approaching, I became keenly aware of all the commotion concerning the Mayan calendar and that something big was supposed to happen on Dec 21st. I read all the differing viewpoints – a lot of them being doomsayers predicting the end of the world. Others proclaimed that an ascension of the human race would take place where we would ascend from a 3D existance to a 5D world. For me some of it made sense but most was a jumble of confusing opinions being promoted as fact. After a while, I came to the conclusion I was simply substituting one belief system with another – both based on looking for answers outside of myself. The problem was, during all this, I never sought answers *within* myself.

That's when I discovered meditation – being able to sit quietly somewhere and calm my mind from all the external noise that surrounded me constantly. I found some good meditation music on the internet and began a regular practice of this.

Cloud Message

During the summer of 2012, I was hanging out with a friend at Ocean Shores along the Washington coastline. While we were driving on the main highway north of there, she announced that we had a visitor with us in the car! There was no one else in the car other than me, who was driving and my friend in the front passenger seat. She is rather psychic being American Indian and proceded to describe what this invisible person was wearing in the back seat. Before long, I knew from the description that she was describing my spouse who had passed away from cancer two years earlier. My friend relayed what the spirit entity was saying to her about myself... things that only Toby would have known. I tried to be logical about this as there was no way that my friend could have known that information. Still influenced by some of my previous religious indoctrination, I expressed my doubts and said that perhaps it was demonic in nature and we were being fooled. However, as we pulled into the beach front area, I then said "Well, I need more proof that this is Toby contacting us". Just then I looked up at the sky through the windshield and beheld what is seen in the photo on the next page.

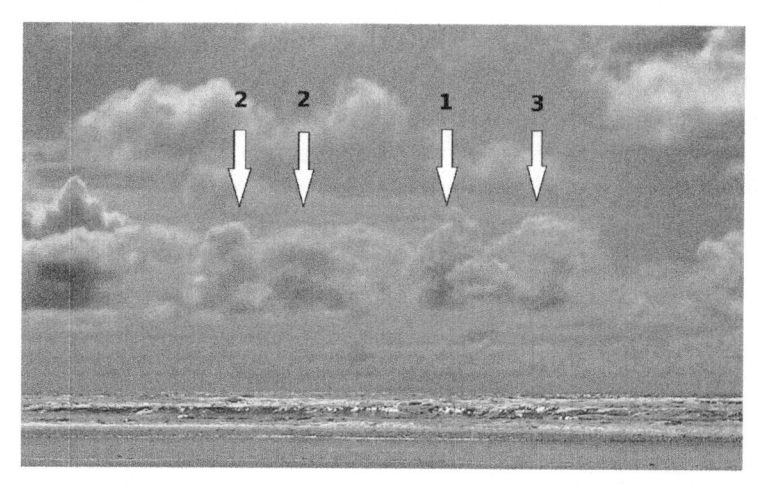

At that instant I could clearly see the numbers "22 13" and immediately took a picture as proof (it's more evident the smaller the photo). Later that evening, at the hotel, I did a Google search for "22 13" and the first thing that came up was a reference to Rev 22:13. I looked it up and the scripture says "I am the Alpha and the Omega, the first and the last, the beginning and the end." It's still hard to relate how I felt but it was a feeling of peace like my life was being guided by forces unseen.

This was the beginning of my exposure to *real* truth – the answers I was seeking as I began my journey through the veil. As I continued my meditation practice, changes started taking place within me and I began to simply *know* things about the nature of the universe, what God really was and more importantly, who we are. As I continued to mediate, the answers came from *within* and not external to my being. I began to appreciate that we are much more than what we were programmed to believe. We are energy/spirit beings clothed with physical bodies. Our energy auras even extend out beyond our bodies as the extention of our actual existance. Meditation allows us to open up our physical self to the part of us that's spirit which has access to all true knowledge.

It was this spiritual awakening that allowed me to pay attention to my own intuition and how it was guiding me. This was the main reason I moved to the mountains – I was spiritually guided there! Part of this awakening revealed to me that we all have what is termed psychic abilities (telepathy, telekinesis, foretelling future events) but for most folks it's dormant due to, once again, belief systems that lock down our minds. In the last few chapters of this book, I delve into why and how the human race has been enslaved by mental conditioning thus creating a veil of separation between us and the real world that the Sasquatch are a part of.

Chapter 2
Meeting Barb Shupe

You can only be in Greenwater for just so long before having to stop in at the only general store in town for supplies or gas. There, on one of my visits to this store, I came across Barb Shupe who worked there as a cashier. I'm not sure now exactly how the conversation came up but I'm sure it had to do with me looking down in the glass display counter and seeing a couple of cast footprints of some very large feet.

Barb Shupe with her Squatching companion Gabby

I asked her "What's this?"

Barb, being very enthusiastic about it, said that they were replica footprints of the Bigfoot. I was intrigued by this though still somewhat skeptical by replying "Really?" I'm not sure what was said next but Barb did give me an earful as she related her personal

[17]

experiences of the Bigfoot she encountered recently while they were running after elk. I'm sure some of my responses were "You mean they're real? I always thought they were just a myth. I mean, I loved watching the movie "Harry and the Hendersons" but it was just fiction!"

Well now, every time I went to the store, I would ask her a little bit more about the specifics of her encounter – what he looked like and some of the sounds that were being heard. She talked about whooping sounds and what sounded like a large stick being knocked against a tree several times. At that point in time, I did not have those experiences, but it kept my interest up since I was definitely open to the possibility.

I told her that I would like to join her sometimes on one of her hikes into the forest but it was not high on my priority list with all the other things going on in my life. However I really wanted to believe that the Bigfoots were real.

I didn't have long to wait…

Chapter 3
Initial Vocalizations and Sightings

Around eleven pm on Sunday of June 23rd, 2013 I decided to call it a night after hanging out at the local tavern. As I started walking across the dark parking lot towards my RV, I heard some loud sounds coming from the woods behind the tavern. Though the calls were similar to those of the Barred Owl, there was something inherently different about these sounds. The repetitious calls were very loud with the end of each call being very deep and gutteral.Unless the owl was at least fifty lbs I would have been hard pressed to believe that such a deep and powerful voice could come from a bird. The sounds alarmed me so that I turned around and went back into the tavern to tell someone. Being very excited I told a few friends "Hey guys! You need to come out here and listen to this!" A couple of them, having just come from the outdoor beer garden, said they had heard it too and were puzzled as to what it was. One patron though didn't think the sound was anything out of the ordinary and said it was just an owl. However, several of us were not convinced that this was something "ordinary" but no one seemed motivated at that point to go back out to investigate further.

I headed back out alone to the parking area and listened for to a couple of more calls before deciding to respond back with my poor imitation of the same call. There seemed to be an immediate reaction as the calls coming from the woods increased and got louder the more I called back. So I had this back and forth rapport going on for quite a while as a pattern soon developed. If I slowed down in making calls it would also slow down. If I stopped calling it also stopped until I called again then there would be an immediate response!

I rushed back in the tavern and was able to convince a couple of them to come outside. We felt the sound was slowly moving towards the beer garden side of the tavern so that's where we went. As we listened, the calls were very loud. We were standing along the back fence to the woods and whatever was doing it was extremely close. The one guy with me also started to do calls and was getting consistent responses. This kept going on for forty minutes before he got tired and headed back in. I, however, was still stoked by whatever this thing was so I followed the sound past the beer garden towards the next street. I walked to the edge of the woods as the calls continued, knowing it had to be less than a hundred yards away up in the trees. The sound then became more distant as it made it's way toward the river to cross over into the main forested area. It had to be well past midnight so I retired for the night.

Barb interviewed me a few days later about this experience and I told her I was completely convinced at that point that what I heard was in fact the Sasquatch. I came to believe they were real and that some of them were living in the woods nearby. I mentioned to her that I was already starting to hike up the same trail she used frequently and felt that, if I tried to use the same call or simply talk out there, then the Sasquatch would know by my voice that it was the same person they were interacting with near the tavern. I felt that the Sasquatch were very timid creatures when it came to humans and the sound of someone familiar would probably help draw them in closer. This way, they could observe me and see I mean them no harm and hopefully encourage them to use more vocalizations. However, what I really was wishing for, was an actual sighting!

As Barb concluded her interview with me, I did mention that a couple of months earlier I had downloaded on my phone some bigfoot sounds posted by other researchers. The next day, after this experience behind the tavern, I ran into the person who stood outside with me for forty minutes that night. I pulled him aside and said "Listen to this!" As I played the sound from my phone, his eyes got real wide and almost fell out of his chair. "Wow! I didn't know you got a recording from last night!" I replied that I did not record that - it was a recording of a Bigfoot I found on the internet. He was blown away by this! It was identical to the sound we heard, giving both of us further confirmation that this was the real deal!

Incidentally Barb has a YouTube channel called "Barb and Gabby" and I believe this was the first time I appeared on her videos. From this point forward, I became more involved with learning about the Sasquatch as well as joining Barb on many of her adventures that continue to this day. Since then, we have become close friends and I have also had the privilege of getting to know her dog Gabby – Barb's partner on her videos. I'm so thankful I met her as she was very instrumental in me getting to the point where I'm able to write a book about my personal experiences and those I shared with Barb and others. I have learned a great deal about the Sasquatch through her and, in the process, introduced me to a great bunch of folks in the Bigfoot community!

Over the course of the next three months, I would occasionally hear those same owl imitations but also some howls and whooping sounds coming from, not only the woods behind the tavern, but also near my property and across the White River. It wasn't taking me very long to draw my own conclusions as to *who* was doing those calls.

On Friday September 6th, 2013 I was once again heading back to my RV at night across the tavern parking lot and heard the same loud sounds but this time coming directly behind two side-by-side buildings positioned along the back of the property, next to the woods. There was a ten foot gap between the buildings that led directly into the woods. This is where the sounds were loudest. I wanted to move in really close to the source, so I gathered a little more courage and, with a small but powerful LED flashlight, started walking towards the darkness between the two buildings. The flashlight was off while I continued to walk back to a clearing that was pitch black as the sounds got very loud… it was as if they were just a few feet away!

I pulled the flashlight out of my pocket, pointed it towards the trees above me, and flipped it on. Immediately, that single loud Barred Owl imitation become sounds coming from at least five different creatures! All sorts of loud sounds were being made including, what I could best describe as, monkey chatter and short whoops. Even the Barred Owl sounds were a bit different and were mixed in there with all the chatter. My gut instinct was, whatever was up in the trees, was now feeling threatened by my presence. I briefly shined my light around for a few seconds more but the sounds got louder and I could tell they were getting closer to me!

The last thing I wanted to do was to turn off my flashlight and have my back towards these guys… especially in the pitch black night. So I just kept the light on towards the trees as I turned around and walked out back to the parking lot.

After a few seconds, the noise finally died down to the single sound and began to move across the trees towards the far side of the parking lot where my RV was parked. While all of this was happening, there were three guys hanging out in the back that started taking an interest in what was going on as they heard the commotion going on behind the buildings. They had a good sized flashlight and followed me towards my RV to get a better look at what was in the trees. As the light was shone up in the tree next to the RV, we didn't spot anything but the chatter started up again and quickly started going back in the opposite direction towards the back of the two buildings. We decided to follow the racket right through the woods until I was able to determine the sound was coming directly above me in a very tall tree. "Quick! Shine the light toward the top!" I insisted. As soon as the light hit the highest branch, I saw for a moment what looked like a four feet diameter ball of grayish/black fur hanging on to the branch. It immediately propelled itself with its feet from the one tree to another at least twenty feet away! (In talking with Barb, she has seen one do the same thing the previous spring.)

The fellow holding the flashlight moved it away from the spot where the creature jumped and I was not able to see it anymore although other creatures were still in the trees making noises. I felt it was best for me to stop pursuing this so I left to get some shut eye for the night. However, I later learned, the three guys that were with me were still excited about what these things were and continued to chase the sounds as they were moving across the treeline into the other street. They finally were able to see five distinct beings moving along in the trees, confirming my estimate of how many I thought there were, although they could only give vague descriptions of their appearance. They did agree on the color being a grayish/black as I described earlier. Once again, the creatures took off down the hill, crossing the river and disappearing in the forest.

I personally learned a valuable lesson from this experience to _never_ shine a flashlight out and up in the trees if I suspect that there are Bigfoots nearby. I now will only shine it towards the ground only to see what's a few feet ahead of me when walking at night.

These vocalizations and brief sightings convinced me that the Bigfoots were real and were coming into our town especially at night. Even one camper next to my RV who claimed to be an unbeliever was shaken up one night by seeing two seven-foot tall Bigfoots walking across the highway next to the tavern and past the back of my RV and the little cabin he was staying in. He was pretty shaken up by this and does not like to talk about it. It took him a little while to get over his fear of staying in his cabin at night.

There is one more incident, though I'm not sure exactly when it took place. I was hanging out up the trail from where Barb lives and, being bored, I decided to "act" a little bit like a Sasquatch. I found a small dead tree, pushed it over and beat my chest like Tarzan while whooping like a Sasquatch. Afterwards, I felt kind of dumb for doing this and started back down the trail. As I came to a bend, I heard a very loud scream calling "Ai-ya-ya-ya-ya" followed by two distinct Barred Owl calls. It was right there not more than fifty feet away from me on my left and as I turned to look, nothing was there. It sounded a lot like Barb's voice, who I thought was messing with me out in the forest. When I arrived at Barb's cabin, I was shocked to find out that it wasn't her after all, as she had been inside watching TV during that entire time! This, I would consider, is one of the best vocalizations I have ever heard but unfortunately, the one I did not record.

Chapter 4
Giant Log Slammed in the Ground

In May of 2013, I was fortunate enough to purchase some property with a small cabin near the White River in Greenwater. For most of that year, I would alternate between living at the rented spot at the tavern where I had the two experiences and at my new place until I secured a water line and fixed up the cabin.

After some weeks, I had cleaned up around the lot and had built a nice fire pit with all the large rocks that were laying around. It was getting late evening and I wanted to build my first fire using the dead branches I had picked up. I had it going pretty good and was standing there gloating about my accomplishment while enjoying the heat from the fire. By now, it was just barely light and I had to go in the RV for a few minutes to take care of some things.

The minute I opened the RV door to step out again, I saw something small by the fire take off towards the woods nearby screaming like a monkey in fear. It happened so fast that it was already long gone by the time I stepped out completely. I *knew* the little visitor had to be a small Sasquatch who was very curious about the glowing fire and took a chance to check it out. It wasn't long before I started hearing whooping calls in those same woods next to my cabin. This occurred especially in the late evenings.

Moving forward to 2014... I was trying to develop a healthy lifestyle by taking daily walks in woods along the White River where the little guy ran towards. One day, I felt compelled to place a triangle symbol on the ground with a rock from river in the middle, in order for me to place some food gifts in hopes that the Sasquatch would venture over here. Hearing Barb talk about her own gifting area success gave me the push to do this.

One thing that happens when you travel the same trail daily is you begin to take note of little changes… and maybe not so little… in the forest, such as snapped tree limbs or fallen trees. I knew a few recent wind storms brought down some of these trees, especially the ones that were already dead but were still standing.

One day in March of 2014, I walked up to my small gifting site and discovered a large log leaning against another tree that had not been there the day before. It was huge and I could not even budge it due to its massive weight! Also, from what I could tell, there were a couple of branches partially attached to the top of this log but they were pointed down, not up. The upper end of this log was actually on the ground with the heavier bottom being leaned against the smaller tree. As I examined the ground end of this log I noticed it was not *on* the ground but *in* it! I came back with a shovel and started digging around the perimeter of the log to see how deep it was driven in the ground.

After digging down a full foot, I soon gave up. What in the world could cause this log to be rammed in the ground this deep? The one logical possibility was that it was part of a larger tree nearby that snapped off up in the air and plunged in the ground and then conveniently fell forward against the smaller tree. However after looking around the general area, there were no trees nearby that matched the necessary criteria. Something had to be large enough to not only pick up that log, but also have the strength to slam it into the ground with enough force to bury it a foot deep!

After bringing Barb out to this spot, we later found a fairly large footprint that was deeply pressed into the ground next to the log. Maybe it wasn't a coincidence that this happened right next to my gifting site – it could have been a clear message to me that the Sasquatch were starting to take notice!

Chapter 5
My Psychic Awakening

During this same time period from 2012 – 2014, I mentioned I was undergoing a spiritual awakening through meditation and looking for answers from within. As this was going on, I started noticing repetitious numbers appearing all over the place. Before I saw the numbers "22 13" in the clouds in the summer of 2012, I was seeing the number 666 everywhere. Now I know what some may think that it was the sign of the devil following me around but it was relentless. I would see licence plates such as "ABC666" with this embedded number all the time. I would order lunch at a restaurant and the check would be $6.66. It is one thing to see this once in a while but everyday and numerous times at that – there was something supernatural going on here! On the internet there was this numerology calculator that would take a first and last name and calculate a three digit number. I tried a number of names (even some of my old bosses' names) to see if I could get 666 to show up but other numbers would show up instead. I tried "Samantha Ritchie" and it also came up with something else. Then I thought "Hmm... Let me try my birth name (I had changed my name back in 2009) and typed it in. As I hit the enter key, I felt a distinct shock in my body when the number 666 stared at me in a bright red color!

Keep in mind I was still holding onto some rigid religious ideas while all this was happening. In the summer of 2012 when I saw the numbers in the clouds, I had mellowed out and was adopting a more spiritual viewpoint about life. Just as quickly as they appeared, the number 666 started to wane from view. However some of the new numbers would make their appearance in my life and have been with me to this very day!

I'm not sure the day when it started but the numbers 44, 444 and 111 started to appear and with each passing day they would appear with greater frequency. As was the case with 666, I was seeing them everywhere – clocks (4:44am/pm), licence plates, restaurant checks, you name it! A strange coincidence was parking next to three other cars that had triple numbers in consecutive order on their licence plates: 222, 333 and 444! I wish I kept a log of all the times and places where this has happened – even passing a dumpster with the number "444444" on it!

Then I noticed that, whenever I drove somewhere, I would arrive at 44 past the hour. This was not something that happened once in a blue moon – it happened every day and several times at that – all day long!

I do have a humorous story related to this that I need to tell. It was during our "Lonesome Lake Canoe Race" held each August for the Greenwater area. It is a fifteen mile drive on a remote forest service road to get there. There is no cell service so I decided to turn off my phone and try to enjoy the races without having to see those numbers appear somewhere. After all, I'm out in the woods far away from technology and devices displaying numbers – right?

Not more than a few minutes after parking my car and walking up to the shelter overlooking the lake, I saw a friend of mine who was taking care of a list of participants who were racing that day. As I approached him, another person happened to walk up to him at the same time and asked what time it was. My friend turned his wrist towards her so she could look at his watch directly. She then says "It's 11:44am. Hmm... that's funny that's the same time exactly an hour ago when I asked – 10:44". My jaw dropped in amazement – I could not escape this, even way out here!

As time passed, it got very intense to the point where I was mentioning it to one of my work associates. Before long, he too was suddenly seeing the number 444 as well. This was NOT some random coincidence – there was something or someone trying to get my attention!

While I meditated I was now starting to know things as if thoughts from elsewhere were merged with my own. Whenever I had one of these sudden thoughts or messages, I would immediately see one of those three magic numbers (usually 44). Since this usually happened while I was driving, it would often be the clock or the odometer beaming the numbers at me. I began to see it as a sign that the thought or message I received was a valid one from what I determined were my spiritual guides.

Here's an example of what I experience on a daily basis in seeing the number 444. As I was driving to town this morning (Feb 9, 2016), I just happened to look down at my odometer and saw the number 444. I pulled over in order to take a picture of it with my cell phone. It wasn't until later when I transferred the photo to my computer that I noticed the cell phone *also* assigned a file name with the number 444 embedded in it. The file name is a date/time stamp and, in the case of this photo, was taken on 02/09/2016 at 9:14:44am – right on the 44th second of the 14th minute. I wasn't even looking at the time when this was taken. Now I have to say, a coincidence is something that happens once in a while, not every single day!

Early in 2014 while attending a spring festival, I came across some crystal pendulums that were used to make spiritual inquiries by psychics. I took one back with me to the cabin and, after doing some meditation, I used it to ask simple questions that would require a yes or no answer. Holding the pendulum steady at the top, if it started to swing back and forth towards me then this indicated a "yes" answer. A side by side motion was "no" and if it started to swing in a circle then that meant "I don't know". I was surprised by the consistent answers I was getting especially when I asked the same question at different times and would get the same answer every time!

Having already gained some spiritual insight as to the nature of the universe and the fact that we are all connected, I asked the kind of questions of an unselfish nature such as "Will mankind survive the horrific problems we're facing in the world?" "Will I play some role in helping others?" The answers were positive for these questions.

However when I asked a very specific question like "Is my role to help save the human race?" the answer came back negative. I asked this same question several times and got the same answer. Now me asking this question was not due to me having some kind of "Savior Complex" but I have always felt my life had a specific purpose to help others and not for the purpose of just catering to my own desires. So it was a logical question to ask to see if my specific role in life was to help the human cause.

This was very puzzling to me until I figured out that I was only asking about "humans" and didn't consider other intelligent beings like the Sasquatch! I'm not sure how the thought came up but I rephrased my question to "Am I to have something to do with the Sasquatch?" The pendulum came back with a strong YES. "Am I to try to contact them and learn from them?" Another affirmative answer came back!

I decided to meditate on this and the answer I received from within was that, in addition to what I have learned up until now, I needed to reach out to the Sasquatch and begin to develop some kind of connection or dialog with them. The purpose was to get insight from them in order for me to help teach others of my kind (humans) their vital message (as you read the final chapters in this book you will come to appreciate what I'm talking about). Also, I am coming to understand that, aside from teaching truths to others, no one person can "save" humanity as a whole. Each and every individual has to make a choice to do what's right with respect to our relationship with each other and that of our planet. Only as a collective group can we make the necessary changes that will ensure our survival as a race in the long run.

Getting answers from the pendulum wasn't the only thing I discovered about it or myself. One day, as I was holding it, I pointed my finger with my other hand at the crystal suspended at the bottom. My finger was a about two inches away from the crystal and as I concentrated, I slowly moved my finger back and forth. I began to "feel" the weight of the crystal as my finger passed directly towards it. Within a few seconds, the crystal started to move and was beginning to follow the movement of my finger! As I continued, it felt like it connected with a tow line to my finger as if the energy was a physical extension of the crystal. It didn't take long before the crystal was swinging wildly... all the while my right hand was holding the chain motionless.

For the next weeks, I was so intrigued by this new ability of mine that I had to show others what I was able to do. I remember having several locals gathered around the fireplace at our tavern watching as I moved my pendulum using this mysterious force of mine! Another new way to move it was to hold the pendulum in front of my head with the crystal right above my eyes in the area where the pineal gland is located. This gland is commonly called the third eye. All I had to do was to concentrate and "push" the crystal with my mind and it would swing wildly within seconds. Also, when I was in Tacoma, I stopped at a tavern where I go regularly and showed this "trick" to several friends who were equally amazed. However, the eyes of one particular patron got real big as the crystal started swinging. "Get that Louisiana VooDoo away from me!" he said as he leaned away a bit fearful.

In March, I had a friend from Australia come for a visit. On his last day in Washington, we went to a hamburger joint in Tacoma and told him about this ability. Sitting at one of those small round tables with two high bar stools, I had him face me from the other side as I pulled out the pendulum. His eyes lit up on seeing this crystal move on its own and insisted it must be some kind of trick.

"Here... let ME hold the pendulum!" He said as he held it in front of me with both hands, planting his elbows firmly on the table forming a solid triangle by which to suspend the chain.

I'm not sure if the pendulum actually moved but when I stared at the crystal and sent a mental burst towards it, he suddenly turned white as a sheet! Being worried I said, "What's wrong? Are you OK?"

He took a few seconds to get his composure before saying, "Each time you were squinting your face, I felt like something solid was hitting my chest!"

I was equally shocked by this – all I could say was "Really??"

Thinking this was just a one time anomaly with the pendulum, I suggested he put it down and simply hold his hands together as before in a triangle and I would try it again. As I stared at the space below his hands, I once again sent out three additional mental bursts.

This time he almost passed out and just about fell out of his chair! He looked as if he had just discovered his friend had some kind of super power and seemed afraid of me! "I felt it hitting my chest again as if you were using your fist!" as he steadied himself again.

We finished lunch and I eventually got him on his way to the airport. Unfortunately between the experiences I was relating about Sasquatch and my demonstration of psychic force, he just dropped out from ever contacting me again. I guess his fear got the best of him and I may have lost a friend in the process.

During the summer, I decided to check out a psychic group that met every week at a local restaurant. On my first visit, I learned that everyone was supposed to leave a small personal item on top of a card with a code word that only you would know. I left a metal neck chain and pendant and chose my personal code word. There were, I believe, six psychics sitting up front and each of them were assigned an even number of personal items, each with its corresponding card and code word. During the meeting, it was the job of the psychic to hold each personal item and make predictions based on the energy they received from it. They then wrote it down on the card.

Later in the meeting, the psychics took turns reading the cards and calling out the code word before reading the prediction. There were quite a few readings that had to be given since there was a decent crowd in attendance. I was very curious as to what they would say for mine but I listened as different readings were given for those before me. All were pretty much the same and involved the usual day to day concerns of relationships and finance. In all honesty, it was like all those relationship-based readings you find in popular magazines…

typical stuff as far as I was concerned. Since the reading was associated with the code word, there was no need to point out any one individual in the group and was done for the sake of privacy and anonymity.

Since I had come in late, I was one of the last to be read. However, something strange happened right before the psychic read my card. He stood up and said, "I need to see who this person is!" I raised my hand as he drew his attention towards me. With a very serious tone, he proceeded on, "I have done a lot of readings for people but this one is very different and I just had to see who I was doing this for." He hesitated before adding, "You have the power of a dragon! Over the course of the next three years many people will depend on you during very trying times!"

This sent chills up my spine as I was singled out in such a vastly different way from all the others. He stood there for a second looking at me and then glanced down at the card and mentioned a couple of other minor things before sitting back down. I was simply blown away by this short reading. I knew I had some psychic power but to compare it or myself to the power of a dragon? Also to what extent will people depend on me? This event just adds to the mystery of all the other things happening in my life. Only time will tell...

In addition to this psychic awakening and based on personal research, it was very apparent that "we are not alone" in the universe. I had come to the conclusion that we are not waiting for Aliens to show up – they are already here and have been all along. I allude to this in a later chapter called "Human's Artificial World". It's also a strange coincidence that, by the time I started to become aware of the Sasquatch being real, I had two different daytime UFO sightings a few months apart.

The first sighting occurred while I was driving back to the mountains from Tacoma. I was traveling eastbound on Highway 410 and was just coming into Enumclaw. On the right side of the highway, there is a strip of grassy land that once was an old railroad line but had been converted into a greenbelt walkway. It was a sunny day and there were numerous people either walking or playing frisbee on the grassy area. As I was approaching the end of this area before entering town, I just happened to look up and saw what looked like a round metallic disk bowed out slight on top and beneath there were three circular areas evenly spaced... a classic description of a UFO saucer!

At first, I thought it was a helium balloons made to look like a UFO. However, it was approximately one hundred feet above the field and was completely stationary. There was nobody with a string nearby that was restraining this thing like one would with a kite. Also, there was little wind and the object was as perfectly still as if it were suspended from an invisible pole. Even though I had a few seconds to watch this and reason on it, I became even more perplexed by the fact that nobody in the field or along the walkway even noticed it! I wish I had stopped to investigate further but there was quite a bit of traffic on the highway behind me and very little shoulder to pull over on, so I kept driving.

The second incident took place a few months later in Tacoma. I had just arrived at my home and as I got out of my car, I happened to look up at the clouds right above the house. I immediately caught sight of a silver sphere that moved behind one of the clouds as if to evade detection. Not to say I was being paranoid, but I did get the distinct feeling I was being watched!

Another strange incident involving time gave me some additional insight into the reality we live in and caused me to question whether or not we were living in some kind of controlled matrix.

It happened while I was driving down a street at night. There was a car approaching from the opposite direction with its headlights on and as it got closer something "shifted" as if I had regressed back in time a few seconds. That same car was again approaching me as if someone hit the rewind button on a movie! I like to also mention that, before this happened, I had synced the digital clock in my car to the actual time as it was five minutes slow. After this event happened I checked my car's clock and it was now several minutes fast! So whatever happened during this "time shift" also affected my vehicle.

What I learned and experienced was not only an awakening of psychic abilities but also a glimpse through the veil of what we call reality. I truly believe that all humans have psychic capabilities though for most it lies dormant. A person's belief system, attitude and, most of all, being in a state of fear will lock down any chance of using these otherwise natural abilities.

All humans use psychic ability and don't realize it. How many times do you know of people calling someone and the other person on the line says, "I was just thinking about you!" or "I was just picking up the phone to call you!" This happens often with someone we have a relationship and with whom we have spent a considerable amount of time.

Case in point is when I called my mother after several weeks of not speaking to her... she often says she was just thinking about me and ready to pick up the phone to call me. The bond we have with someone is not just an expression... it's a psychic connection that stays strong regardless of the distance separating us!

Having these abilities has personally helped me in my quest to understand the Sasquatch and has allowed me to gain the knowledge shared in this book. In the process, I have also learned more about ourselves and the nature of what we call reality.

Chapter 6
The Cloaking Bigfoot

On September 11th, 2014 during the second Barb and Gabby Campout (my first), we hiked as a small group along a small trail leading toward a gifting spot maintained by our friend Sandy. Others from the main group, including Barb's dad and his wife Juanita decided to take the main trail up a little bit further to sit and rest. Along the way, Barb stopped to show us a snapped tree that had recently been broken in the middle while Sandy continued on by herself towards her gifting area. Barb's Ipod was on record while pointing out that something had hit that tree hard and fast to cause the new break without completely snapping off the older break.

While Barb was talking, she immediately caught sight of a domed headed creature running towards the left and away, mowing over the weeds as it went. Barb pointed her Ipod in the general direction and shouts, "What the heck was that? Did you see that? There's something taking off right there!"

According to Larry, one of the members of the group, I reacted to the sound it was making because I immediately turned to look myself but whatever was there had already taken off. Barb continued "Did you see that? I just saw something small and black take off that way fast! Boy I hope it was on camera!".

After a short discussion concerning what was just seen, we decided to go back to the main trail and meet with the others. They were only a hundred yards away and maybe they saw something too. Juanita told Barb that Chita (Juanita's small dog she was holding) starting barking at something that was crossing the main trail. Juanita then looked in the direction where we were and saw something small and black shimmering down a tree and take off at the same time that Barb spotted the larger Bigfoot. Afterwards, we continued on up the main trail to a bridge on a logging road.

In the meantime, Sandy had just finished checking out her gifting spot and was wondering where everybody went. She came back along the small trail and then turned ascend the main trail where we had already been. As Sandy got to the same spot where Juanita had her sighting, she began to hear a woman's voice. Although Sandy could not make out what was being said, it conveyed the feeling of a mother calling in her children.

Later that evening, while we were cooking supper, Barb took the opportunity to review the video on her Ipod to see what was caught. As she got to the part where she had seen something move and run, there was something strange dropping down off of a branch. It wasn't black at all like she expected but whatever it was, it looked strangely distorted. "What the heck is that?" she wondered. She then decided to upload the video to her Windows Surface computer tablet in order to get a larger screen view of what was going on.

"Oh my God! Oh my God!" was the next thing we heard as she was sitting in her chair stunned!

A few of us rushed over not knowing what to think and asked "What's wrong, Barb?"

She just pointed to her computer tablet saying, "The Ipod can't do this! This is Hollywood stuff!" pointing to the spot on the video for us to observe carefully. The tablet screen was still too small to see clearly so we uploaded the video to a laptop.

We took turns looking at the laptop of this strange distortion jumping down from a branch and taking off! Everyone suggested we needed to see this on yet a bigger screen. Barb's dad and his wife Juanita were camping in their RV in a campground just above us and had brought with them a special computer that was built into a twenty four inch screen.

As I reviewed the video, I could tell that there was a creature jumping down but it had no real color. Instead it seemed to match the colors *behind* it as it was moving. It was as if the creature was translucent – being able to see through it but still having it partially visible as if looking through a soap bubble. Several of us immediately thought of the movie "Predator" when the creature became partially invisible and jumped down from a tree. It was identical to this except this was NOT science fiction! This was for real!

We later did a re-creation of events to determine where exactly the anomaly was and where it ran. We determined it to be seventy five feet away from our position and found two sets of nine and one half inch footprints… one going to the left and into a cubbyhole we found under a log and the other was straight back and around closer to where the other group was at.

The video generated a lot of interest in the Bigfoot community and several researchers produced their own analysis videos of what they thought this creature jumping down from that branch was and why it seemed to be cloaked (or gave the appearance of such). The opinions ranged from it being a Grey Alien, a Bigfoot and even someone suggesting it was simply a leaf falling to the ground. The large dark spot seen in the original video did, in fact, look much like the large black eye of a Grey Alien.

I did my own analysis on the video and personally determined it to be a larger Sasquatch stooping down before running off with a smaller juvenile on its shoulders. The large eye I saw earlier was actually the head of the juvenile lunging forward from the jolt of the larger one stooping down quickly. My analysis video can be found on the Planet Sasquatch YouTube channel and is called "Episode 13 – Cloaking Bigfoot Revisited!"

Chapter 7
The Light Portal & Orb

On the following day (Friday September 12th, 2014) I made a mental intention to return to that trail and not bring any recording equipment whatsoever. My reasoning was, since the Sasquatch seem to know when electronic equipment is around and do what they can to shy away from it, it was more important for me to have a personal experience of some kind than to try to capture it on video. Actually, I was kind of hoping for one of them to walk right up to me and, at the very least, exchange glances.

As I walked up the trail, I noted that the wind was picking up and upon passing to the other side of the little valley we now call the "field of dreams", a strange but euphoric feeling came over me. I began to talk to everything around me, the trees, the animals, the birds and especially the Sasquatches. I was up there alone and within a hundred yards from where the cloaked Bigfoot mentioned in the previous chapter appeared. I turned to go back and encountered an even more amazing sight! Suspended in the air a couple of feet above the ground was four rods of glowing reddish/orange lights forming the shape of a rectangular box. This wasn't there a few minutes earlier and must have appeared after I walked past it. [Shown is an artist conception of the light rods over the actual spot in the photo]

At first I thought that it was strong sunlight reflecting off the ground until, as I got a little closer, noticed that the light rods were above the ground and not reflecting off of anything! As I moved around it, the rods were stationary as if they were part of a 3D object on a CAD software program. Also if anyone has watched the "Tron" movies, this was like seeing a virtual light object in real life.

A strange calmness came over me as I was not in fear of this but was awestruck and said to myself, "Well, you don't see that everyday!" Did I come upon a portal of some kind and was this associated with the Sasquatch in some way? I didn't dare get closer to it though as I really didn't like the prospect of being thrust into some other dimension or world. I kept my distance, being no closer than fifteen feet from it, and felt it was best to just observe especially if I had no understanding of what I was really dealing with!

As I passed it and looked back, a large beach ball sized orb appeared out of nowhere. It was translucent like a giant soap bubble and glowed with orange electrical sparks of energy. It slowly made it's way between and into the four light rods and within seconds, the whole thing disappeared.

On September 28th, 2014 I came back to look for any evidence that the portal may have left behind. As I walked to the same spot, I looked down and discovered the ground area where the portal had been suspended was completed sheered away and formed a nearly perfect triangle (see photo on the next page).

It was if someone took a large sharp blade and scraped the leaves and undergrowth clean away. As I examined it closer, I even noticed that all the trees roots that were close to the surface were also shaved clean to the same level as the ground. It's one thing for someone to simply clear off an area down to the soil but it's not possible for someone to precisely shave across a perfectly flat area cutting through not only dirt but also rocks and roots!

What could possibly cause this? All I can do is speculate but since the light portal was over that spot, it's possible that whatever was within it's energy field (including part of the ground) also disappeared with it as it pulsed out of sight! This experience still affects me each and every day. It was like watching the special effects in a Sci-Fi movie but this was for real! The unseen reality that most people never experience was coming at me like a freight train and this was only the beginning as I would soon find out within that hour!

Chapter 8
The Local Clan Observes Me

This now brings me back to September 28th, 2014 and just a few minutes after I took the photo of the cleared triangle on the ground at 5:23pm. I was still hanging around in the "field of dreams" area and continued to take pictures with my cell phone camera. About twenty minutes later I decided to sit on a log next to the trail to rest. As I sat there I quickly dozed off into a deep trance-like sleep. I finally slowly woke up but was completely disoriented and, for a few seconds, did not know where I was. I felt like I was sleeping for a couple of hours even though only ten minutes had passed!

At 5:53pm I had this strange urge to take more pictures with the cell phone but try to catch only what was going on *behind* me. I remembered hearing stories of how the Sasquatch would be behind people and sometimes they would catch a glimpse of them using a cam attached to their backpacks. I switched the cell phone to use the front "selfie" camera (or so I thought!) instead of the back. I held it above my head I clicked away a number of times hoping to see a Bigfoot right behind me.

As I sat there, I did a quick review of the pictures on my phone and discovered that I had the opposite camera active and was actually taking pictures in *front* of me instead of behind me. There was one decent picture of the landscape in front of me but due to the small screen on my phone I didn't notice anything out of the ordinary. I lingered a while longer as the sun was setting and shining through the trees before heading back to my cabin. I think a few days went by before I loaded the photos onto my laptop and began to take a closer look using the Windows photo viewer. This allowed me to zoom into small areas of the photo and see in greater detail what was actually there.

The photo shown at the beginning of this chapter is one of several that I was examining. After enlarging the middle portion of the photo, I couldn't believe what I was seeing!

The local Sasquatch clan within the circled areas

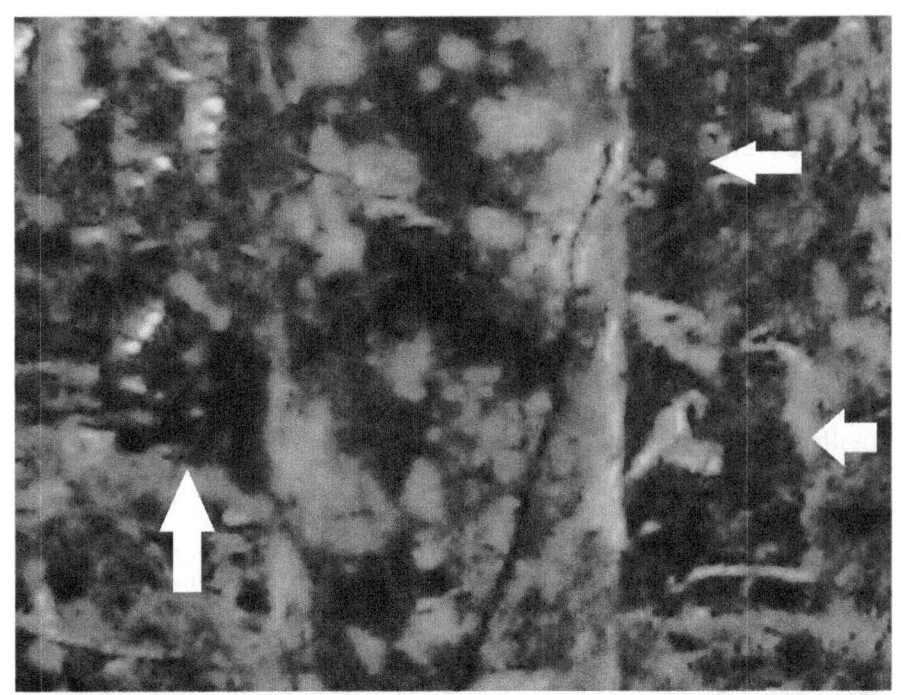

The upper arrow points out a very human-faced Sasquatch on the side of the tree in the middle. The lower right arrow points to a very young one in a sitting poise while on the left, there's another adult with red piercing eyes (noticable on the color photo version)

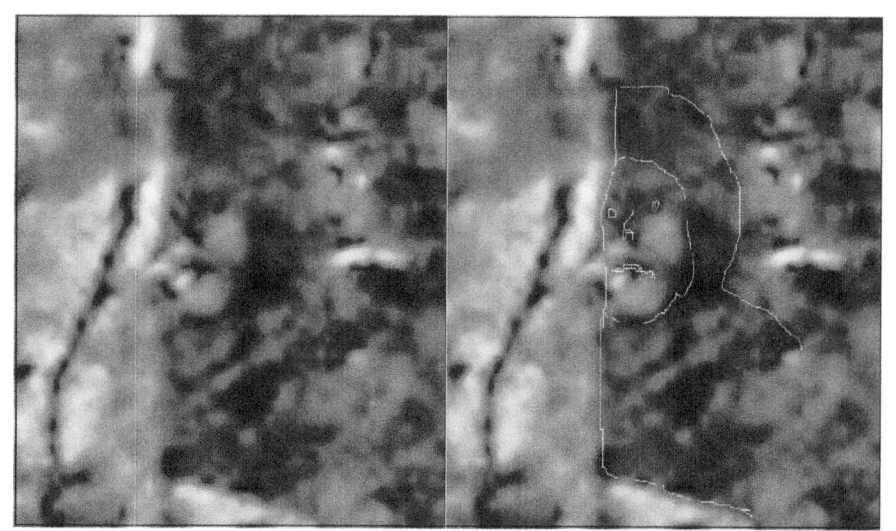

The human faced Sasquatch with its face partially behind the tree

The young one in the middle and a dog-like creature to its right

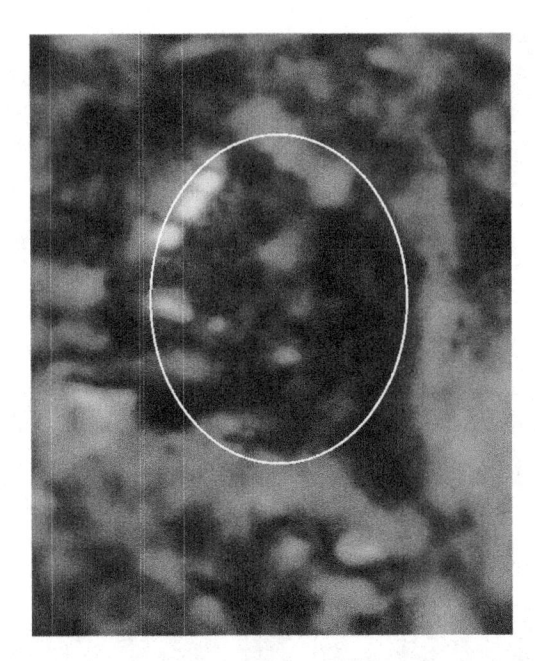

The above being was facing sideways while holding a little one

**This adult was further back and to the right of the others and is hidden behind the bush except for the upper head and eyes
(See other circle in main photo)**

Though the photo was a bit dark and grainy, I could easily make out a number of different Sasquatch... all standing and looking at me as I took those pictures! Just to make sure of what I was seeing, I took additional photos of the same spot and found the absence of the Sasquatch in those pictures. On my YouTube channel, Planet Sasquatch, you will find a video called "Episode 18 – Meet The Family!" covering a breakdown of the original photo.

Why did they allow me to take their picture so openly? I had never in prior research seen any photo like this on the internet from other Bigfoot researchers! Did they somehow read my mind and felt convinced that I was taking a picture in the opposite direction as I thought and didn't worry about hiding? What attracted them there in the first place?

Well I *do* snore... very badly... when I'm sleeping. Maybe they heard me and were concerned about these strange noises coming from me. Perhaps like an animal in pain, waiting until I woke up to make sure I was ok. Whatever the case may have been, one thing is for sure - they did not try to harm me while I was sleeping. This first photo of Sasquatches was just the beginning of what I would experience in the following months.

Up to this point of discovering that Bigfoots can cloak and light portals do exist, I knew my life was changed forever and that we were not alone as intelligent beings in the universe as mainstream science and religion had led us to believe. I would now spend as much time up in this area with hopes that I would come across something again - Sasquatches, portals, other unknown creatures - you name it!

Subsequently, different members of this clan did, in fact, start showing up in my successive photos further confirming their trust in me and the fact of them not being intimidated by my presence.

Chapter 9
Direct Sasquatch Interaction

On October 28th, 2014 I took one of my regular walks up the trail to the "field of dreams" area. It's a very unusual spot where there is a natural clearing in the middle of the forest. Previously fallen Cottonwood trees seem to be placed in such as way as to form glyph structures on the ground. This becomes apparent if you were to somehow look down from certain height. A creek runs along one side of the trail and a marshy wetland area on the other. Next to the marsh is a ridge with very steep sides that makes it difficult to hike.

We have heard a lot of activity up on that ridge... like tree knocks on multiple occasions. Barb and I have climbed up there on occasion and have found a number of peanut jar lids along the side of the trail leading up to the top. Our best conclusion was that the younger ones were taking the sealed peanut butter jars from the gifting spots in the valley and retreating to the top of the ridge before eating. From a strategic standpoint, the ridge is the perfect place for the Sasquatch to hang out as they can observe the entire valley, knowing that, if there was any sign of danger, they could easily move across the ridge out of harm's way.

On this particular autumn day, the sun was out, brightly lighting the little valley. On the other hand, the trees on the ridge provided contrasting shadows though one could still see through it though it was darker. As I entered the valley, I looked around a bit before taking a couple of photos near the trail and toward the creek.

I then turned my attention to the edge of the marsh where it met the bottom of the ridge. As I was looking at the shadows, I noticed the area between two of the trees was darker than the surrounding shadows. What was stranger, other than the fact that I could not see through it, was that I saw this dark area moving in a slight wavy motion. It also reminded me of a dark moving mirage usually seen in the desert on a very hot day.

Intuitively, I *knew* this had to be a large Sasquatch so I took out my cell phone, waved at it and took the following photo.

First photo taken with my cell phone camera at 5:05pm

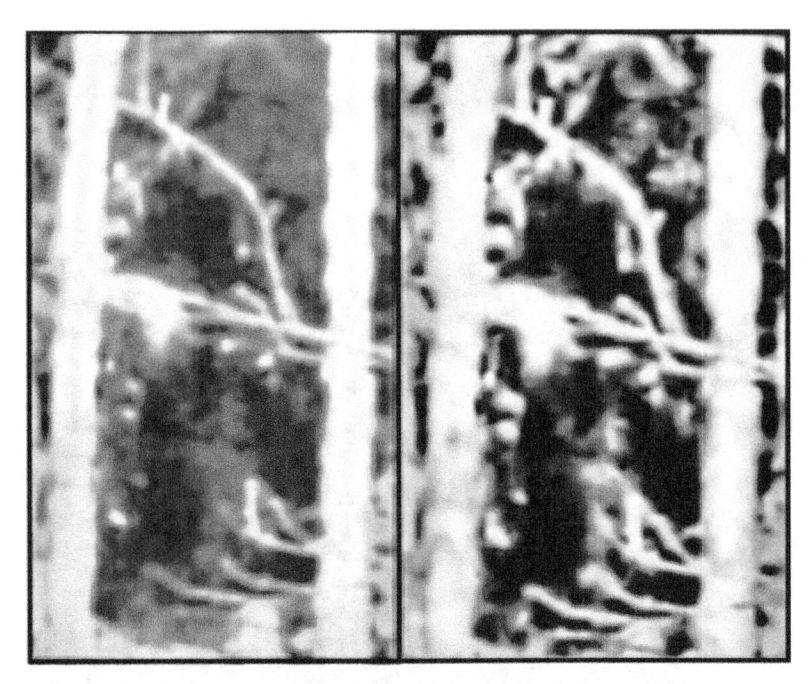

Closer enlargement and with extreme contrasting added

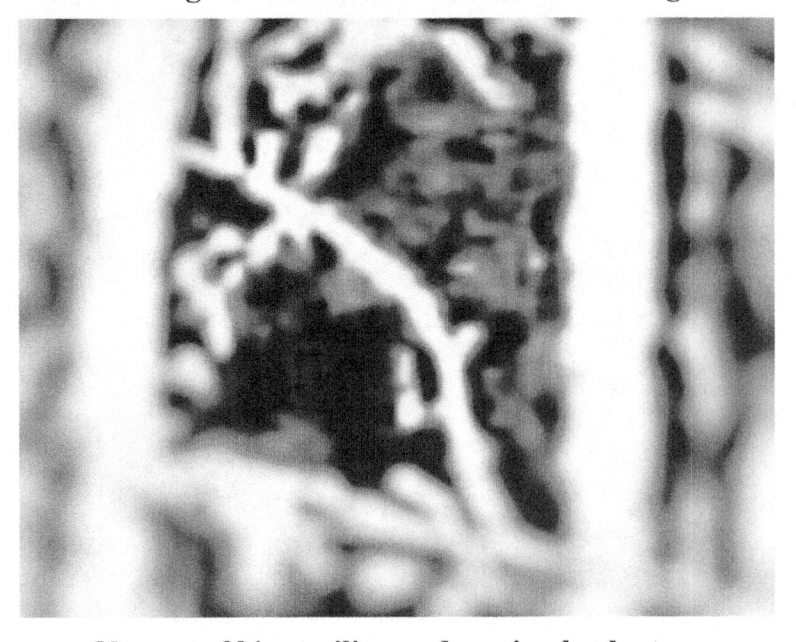

Closeup of him smiling and waving back at me

Honestly, I was surprised that I could determine it was a Sasquatch in the first place! The very dark, smoke-like film that shrouded it was definitely one of the ways they're able to hide themselves especially when humans are nearby.

I believe it really caught this guy off guard when I started looking right in his direction and began waving. Based on the smile I see in the first photo, I guess he was very impressed by this and waved back!

Afterwards, I walked up a little closer and took the second photo shown below:

His body is more visible in this photo, though it's still shrouded in a dark, smokey film. It required some basic contrast and lighting adjustments in order to show the unmistakable outline of his large physique. It may not be as evident here but, on closer examination of the photo, he is actually pointing in my direction with his right hand and has his head turned to his left. Mostly hidden on his left, was another Sasquatch peeping out behind a tree. It doesn't take much logic to figure out that I'm the topic of conversation between them, which probably went something like this, "Check this out! She can actually see us!"

To this day, I keep a framed picture of the Sasquatch waving at me as a constant reminder of their impact on my life. I can only hope that someday I will have the opportunity to meet this same fella and perhaps sit down for a little conversation!

Chapter 10
Backyard Encounter

On March 22, 2015 I had an experience as I was returning to my cabin after dark. As I pulled up in my car, there was a small herd of elk in my yard so I got out and shooed them away. I then walked back to the car near some large cottonwood logs that had been laying there for years. At the other end of the logs is a small strip of woods along the back of my property separating my yard from the neighbor's backyard. As I looked in that direction, I immediately caught sight of different lights including a piercing set of reddish/yellow eyes moving within the trees.

I thought that maybe there were still some elk hanging out in that wooded strip that were part of the herd I just shooed away. However, I got to thinking that this may not be elk because their eyes would normally reflect back a greenish/white color when the car headlight shined on them. In this case, there was no light shining towards the woods - the moving eyes were glowing on their own! I also remembered this same reddish/yellow eye glow from a previous experience where I was in the forest after dark and encountered the eyes of the Sasquatch just down the trail toward where I was heading.

I kept looking from my position at the end of the logs while trying to figure out not only what the stationary lights were but also the moving set of red eyes. Occasionally, while I was looking, the eyes would seem to swing to the left and disappear for a few seconds before reappearing by swinging to the right. This was right next to the stationary lights that seems to flicker with different colors. As I intensely watched the lights, it dawned on me that this was coming from the neighbor's back window to their family room. The flickering colorful lights turned out to be their TV. My neighbor has a five foot wooden privacy fence along his side of the wooded strip that encloses his backyard and the house window (where the TV was visible) was well above this fence and visible through the trees. This means the eyes I was seeing were at least a couple of feet higher than the fence.

I came to the conclusion that it is was a Sasquatch hanging out in the woods watching my neighbor's TV! Standing at the far end of the logs I started speaking at him "Hey, how are you doing? You're more than welcome to visit here whenever you like!" It was very dark and though I could not see his body, I did see his eyes swing back towards me and stayed fixed in my direction. They didn't move at that point for a good minute but did blink a few times. I continued to talk to him, though I don't remember the exact dialog. I was simply extending my friendship to him. Afterward, and while I was speaking, I could see those reddish/yellow eyes move across the strip of woods glancing at me every couple of seconds.

The eyes were now facing me at the opposite end of the logs. At this point, I couldn't help but notice the spacing between his two eyes – they had to have been at least six to eight inches. Though I couldn't see his head or body it definitely had to have been pretty big! I thought he may have been around seven feet tall as he was standing just inside the treeline. However, he started to move slowly toward me and all I could see were these glowing red eyes beginning to tower over me. Try around eight to ten feet… I definitely had to look up to him!

I thought I had the courage to stand firm and meet him directly. Most people in this situation would have either passed out or have been frightened out of their wits! I felt confident in my ability to not be fearful but, unfortunately, this was more than I could take. Though I felt some fear, I quickly came up with a ready excuse to leave and said, "Sorry, I have to go now to take care of some things... but I'll be back later if you're still around..." I turned around and paced back to the car, got in and left. When I returned, about an hour later, he was gone. Talk about a missed opportunity for, at least, a good hand shake or a hug! I still could kick myself over not taking it to the next level of contact but obviously I was not ready.

The next day, I examined the back area and discovered a lot of foot activity within the strip of woods. Aside from one small distinct footprint, the rest of the ground was fairly well-trodden. If this were elk in that small area then there would have been plenty of dung in evidence as there is in my yard but not in the woods.

The Sasquatch certainly had my number as this was not the first time they came to my cabin for a visit. I had previously mentioned that I had seen these same eyes glowing on the trail. A few weeks earlier, I had stayed out late near the "field of dreams" recording howls just down from me. It was already dark and I was making my way down the trail until I saw two sets of eyes staring from down the trail... between me and where my car was parked. It was a staring contest for a minute until it howled at me. I stood up and howled back which I think was something they didn't expect! I continued on down to my car and went back to my cabin. Some weeks later, I was leaving my cabin while dark outside. I had just locked the door and before I could turn around, I heard the loudest howl right behind me causing me to jump right out of my skin! As I turned around to look, I got howled at again although I couldn't see anything in the dark. I then did a soft howl back and it did likewise before taking off back across the road towards their area. I understood what happened though: it was payback time for me. If I get them shook up by howling on their turf then they will return the favor by howling on mine! I do believe it was a good spirited practical joke on their part to teach me a lesson.

This would not be the end of their visits to my place as we will find in a later chapter.

Chapter 11
A Night Walk to Remember!

On April 28, 2015 I arranged to meet Barb late in the evening to take the trail towards the field of dreams area. It was already getting dark and boy, were we in for some surprises! As we approached the small valley in the middle of the forest, we looked up at the ridge and saw numerous eyes shining in the trees. Keep in mind this wasn't any kind of reflection since we left the flashlights off. I proceeded to play the flute/recorder for a few minutes and things really began to happen! First, something was thrown off the top of the ridge and came crashing down below. Some time later, as we sat there on a log, we heard the nearby trees cracking back and forth but not all at the same time... one would crack followed by another... it was very bizarre! In between all this, we could hear the faint whistles in the background mimicking the tune I was playing on the flute earlier. Lastly, as we got up to leave, something large jumped down off a low branch and hit the ground running although we could not see anything. Barb did hear a distinct grunt as we said "Good bye". We also had the thermal camera with us but could not see anything unusual, not even elk, while this was happening. We were definitely looking forward to future night walks just for these experiences!

Chapter 12
Barb and Gabby June 2015 Campout

On our first night (Wednesday) of camping in a remote part of the Cascades, Barb and Gabby along with Sandy and one other member of our group, went for a night walk down the hill from our camp. During the entire campout, we were hearing whistles every night but nothing compared to what all of us were about to experience! We had just finished playing some musical instruments around the campfire when several whistles started. Perhaps they wanted an encore performance! I whistled and received a whistle back each time!

As soon as the group left the camp, they heard faint calls. After a few minutes of walking, they stepped off the road into what they thought was a campsite they visited during the day. Using flashlights to enable them to look around, Barb expressed doubt that this was the same campsite as the creek was no where to be seen or heard. While they were still discussing where the campsite might be located, there was a ear piercing scream followed by two Barred Owl type calls.

The group stood frozen for a second by the sound as Sandy whispered, "Is it on there?" referring to whether or not Barb had her Ipod on record mode.

Barb whispered back, "That was freakin awesome!"

Even Gabby stood on high alert! One other member of the group who was a few feet away assumed that the scream was being made by Barb. Walking up to Barb and Sandy he said, "You mean that was real?" Barb and Sandy tried to assure him that it wasn't them who made the screaming call.

He was still in disbelief asking, "Are you serious?" Both Barb and Sandy broke out in laughter not only in response to the awesome scream but also from his thinking that they were pulling a fast one on him.

Barb continued to laugh, "I couldn't do that even if I tried!"

He finally said, "So... do you think that was one of our forest friends?"

Barb reassured him, "Yes, I'm sure it was!"

They then determined that the scream came from across the road towards our camp as they continued walking down the road discussing what they heard. Barb stated that could *not* be an owl based on the sheer volume of the scream and amplitude of the sound.

I had stayed at the camp while this transpired. Just prior to the scream they heard, a couple of whooping calls near me prompted me to also turn on my audio recorder while I was sitting by the campfire. That's when I heard the same scream but it was *below* the camp and not as loud as what they had heard or recorded. However, there was a second scream (without the addition owl calls) that took place a minute and a half after the first. This time, it was just down the side road leading to our camp and it was much louder from my position.

Barb and the group returned from their walk to tell me about the scream they heard. I told them I also heard the scream *and* had it recorded. I asked them if they heard the second scream while they were down there. It turned out that they could not hear anything but the first scream. All of us were so glad that we not only had one recording but two separate recordings from different locations to further validate what we were hearing!

Since the second scream came from just past the long line of parked cars and close to the main road, we decided to see if there were any footprints. Early the following morning, we walked across the road to investigate. There was a small ditch and sure enough we found three footprints in the soft mud next to the drainage pipe, all of which measured thirteen inches in length. Sandy selected the best footprint to do a cast that clearly showed the shape of the foot including the toes.

We also walked down to the spot where the first scream was heard and discovered that it was a spur logging road, instead of a camp, that led down to the river. One large footprint and possibly a small one were also found nearby and a cast was made. We also found numerous smaller footprints of about five inches length that had to be from the younger juveniles. On most nights while we were camping, there were several of our group that reported hearing small footsteps around their tents along with a kind of soft jibberish sound.

On the following night, the group took another night walk and this time one of our campers brought his mandolin to play when they stopped to rest by the trail. A foul strong odor of sewage hit Barb and then the others. The camper belting out a pleasant tune on his mandolin stopped, "Maybe they don't like my music!"

It was strange that it was not a consistent smell but would come and go while they were standing there. This was very unusual as there were no treatment plants or restroom facilities within many, many miles. Also, on all the other hikes through the area including the one where the screams were heard, there was no odor detected. While investigating the area where the smell occurred, a foot print and hand print were found just over the edge and into the creek.

Other researchers have noted that, in many cases where the Bigfoots were nearby, there would be a foul odor like you would find at a garbage landfill. However a nearby resident stated that whenever the Bigfoot comes to visit her place, she would smell the strong odor of sewage so it probably depends on the area.

On Saturday, our last night of camping, I had a personal experience in the early morning hours while still dark. I went to sleep in my small tent around 12:30am Sunday morning. There is a small pocket on the inside of the tent into which I placed my running small audio recorder to capture everything while sleeping. Unfortunately, I have the problem of having to get up every couple of hours to take care of "personal business" especially if I drink too many fluids.

At 3:30am nature was calling me badly and I knew I had to get out of my tent. I was a bit intimidated by the dark night and could only see up to six feet in front of me without a flashlight. At first, I entertained the idea of walking across to the restroom hut on the other edge of our camp. As it was, I made sure I had set my tent up near the firepit and safely within camp with the other tents around me on the perimeter. I knew the younger Sasquatch were out there the few nights before making noises and messing with a few of our tents. However, it was quiet outside and I wasn't up to having to deal with a confrontation under these circumstances. Maybe I could in the daytime but certainly not on a pitch black night!) So I unzipped the tent and scooted out to stand up outside.

It was real quiet and no one was stirring so I stepped to the side of my tent to relieve myself. I'm half-way done when all of a sudden, I hear a thud like something jumping down from a tree and then a stomping gallop toward me. That's when I see this large black dog-sized creature run right in front of me and then around the firepit!

My first reaction was, this was Gabby (Barb's dog). Gabby was with Barb in her tent and I figured Gabby may have heard me go outside and left the tent to see me. However, what was so strange about this, was I kept saying "Gabby! Come here!" and she wouldn't come to me. Now Gabby always comes to me when I call her and, as a matter of fact, when I was relating this experience on one of Barb's videos, Gabby did just that after I said "Gabby! Come here!"

I was getting very perplexed with her not coming to me until I said "Gabby! What's wrong with you?" The creature stopped for a second and peeked its head up to look at me while making a strange whimpering sound like a dog does when being scolded. I finally realized that this was not Gabby after all! Before I had a chance to reassess the situation, the creature took off while stomping away. "Oh well..." were my closing words as I finished up and got back in my tent.

I woke up with the light of day and grabbed my recorder to see if it picked up any this action. Sure enough, I found the spot of the recording as I'm waking up and getting out of the tent. What I described above was actually picked by the recording including the stomping, galloping and the whimper. The thump was especially loud when the creature "dropped" in. However, there were also other sounds of heavy grunting or breathing that seemed to be picked up from behind me where as the recorder was set up in the back of my tent.

The best guess was that I witnessed a juvenile Sasquatch running in front of me. I also believe that there was a large adult directly behind me watching all of this and the grunt may have been directed at the younger Sasquatch and, perhaps, that's why it whimpered? I guess if I had realized immediately that I was surrounded by Sasquatches instead of thinking Gabby was there then my reaction would have been closer to passing out!

I also talked with Barb that next morning and she reassured me that Gabby had been with her the entire night in her tent. It was also zipped completely closed so there would have been no way that it was Gabby I saw earlier.

Some may argue that it was a black bear that ran in front of me but it's the sheer speed of what I witnessed running around the pit a number of times before disappearing in the night that discounts that theory. Also, although we stored the camp food in containers, the garbage bags were still open and yet remained untouched, which would have been a different story if bears were present. Further, if it was a small bear, its first reaction on seeing me emerge from the tent would be to have either taken off or the very slim chance that it would have attacked me... neither of these scenaria happened.

Based on my prior experiences with the Sasquatch, especially the younger ones, they are very playful and great at being practical jokers. Seeing me leave the tent by myself was a perfect opportunity to spoof me by "dropping in" for a visit. Remember in an earlier chapter I mentioned how I hid in the woods past dark while they were expecting me to leave at the end of the trail and how I popped up and howled at them when I saw their eye shine. Well they later spoofed me at my cabin as I was leaving in the dark and I have a suspicion that this may have been the same young fellow playing with me again!

These were the highlights of our camping trip in June however it would not be the end of the experiences we were to have on our next camping trip to the Blue Mountains of Washington in the following month!

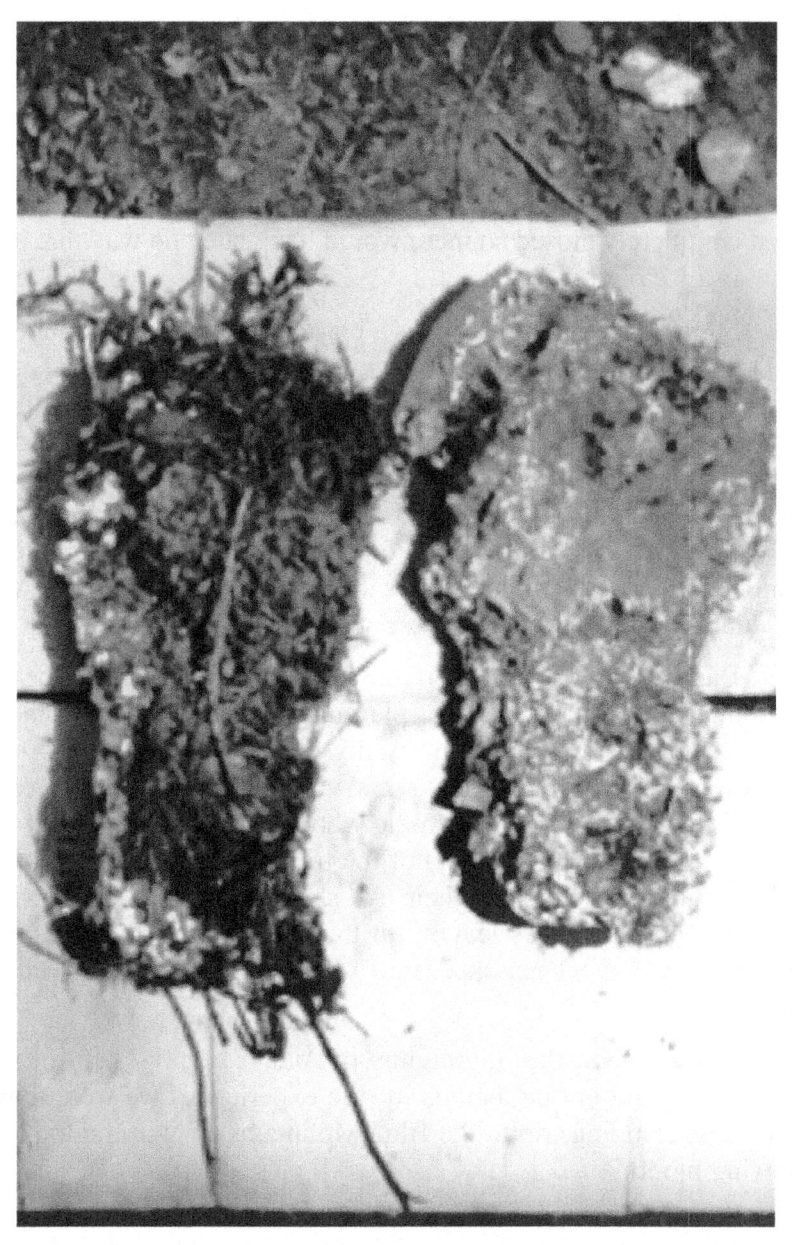

The cast on the left came from down the hill, near where the first calls were heard. The one on the right was from across the road from camp where the second scream was heard.

[62]

Chapter 13
The Blue Mountains Campout

In July of 2015, I drove with Barb to the Blue Mountains in Southeast Washington to attend the Bigfoot Community group campout. The weather on that side of Washington was very arid and hot and, as we passed through several small towns in the low lying valleys on our way, I was concerned that it would be a toasty experience camping outside. However, after we made our turn off from the last town up the road to our destination, I was pleasantly surprised by the dropping temperature until it was a comfortable seventy degrees Farenheit by the time we arrived at camp. It was already 6 pm so we didn't waste any time getting our tents pitched including setting up an outdoor shower hut using a privacy tent (normally used for a latrine), a black plastic shower bag and a rope to suspend it between two trees. Once done, we had time to meet the others who had already arrived and discussed our plans for that evening.

After dark, five of us (Barb, Robin, Schelli, myself and Robin's dog Roo) took off for our first night walk. As we were leaving, Barb clearly saw large, red eyes watching from the shadows behind camp. Little did we know, this was going to be the start of a long list of amazing experiences to be had by all!

We stayed on the road and relied on starlight alone to guide us. Several others mentioned that they too were seeing eye shine and hearing movement nearby. Suddenly, Roo started growling as something was being thrown at us. There was a short pause until another pebble or stick was heard and we knew for sure the Sasquatch were there! We started whistling and laughing, letting them know we *knew* they were there and maybe draw them in a little closer. The eye shine and noise, from things being thrown, seemed to increase and was all around us. Stick snaps were also being heard as the beaming red eyes continued to blink. Someone had a harmonica and between that and whistling, Roo started growling even more. Even Barb swore she saw a black shadow move between two trees. Robin thought they were testing us to see how we'd react. I guess, if it was a different group of people that weren't Squatchers, they'd be running out of there as fast as possible! The activity of the eye shine, their movement, things being thrown and even whistles back at us, continued until we walked back to camp. It was more than we expected for a first night and was well worth the walk!

Late that night, after everyone went to bed, Barb and I decided to sit out on our folding chairs for time. It was pretty cold by then so we bundled up in blankets. We were at the bottom of a sloping open field that led up to the ridge where drinking water was available from a spring. As we faced the uphill slope without trees blocking our view, we were taken back by the brilliance of the stars in the sky overhead. Being in a remote place away from city lights, the stars shined brighter than I have ever seen... it was absolutely beautiful!

During the time we were sitting there we kept hearing small sticks snapping to the right of us and further up in the field. This happened as well as movement and some small whistles. We had made the decision, earlier, to turn off our lantern just in case some young Sasquatches wanted to come in a little closer. It was getting too cold for just one blanket so I got up and returned with my sleeping bag. Being wrapped like a cocoon, I sat in my chair very still while we whispered about what was transpiring in the field. I believe we sat there for a full half hour until Barb whispered that she was going to walk up to the outdoor restroom over to our left then call it a night.

Not more than a minute passed after she left, I moved my head slightly from being motionless when suddenly something jumped up to our right from behind some bushes and started galloping on all fours right in front of me! It wasn't an elk or deer but rather a small black creature about the size of Gabby.

I thought to myself... "Not again! This is just like a replay of that night during the Barb and Gabby Campout the month before!" This time, before it had a chance to run away, I quickly and calmly said, "Hey, you don't have run off. There's no reason to be scared of me." It stopped right in its tracks as I talked to it for at least fifteen seconds before it bolted up the field towards the ridge. Well, that's enough excitement for tonight, I thought as I also retired for the night.

The next morning Barb and I were both looking for evidence

of the younger Sasquatch, so we first checked out the area behind the bushes. We did find flattened spots in the grass as well as some traces of a path that coincided with what I saw. Afterward we were walking over to my car and noticed some small fingerprints initially on the side facing the main camp. My car was parked faced in the opposite direction the previous evening and night prior to these pictures being taken the next morning. As we walked to the opposite side, the right windows of the car were covered in prints as the photos show! Having driven to this spot twenty-five miles on a gravel road the day before, my windows were covered in dust and pollen. This made it easy to spot all the different prints we were finding. Since my car was parked on the edge of camp, we came to the conclusion that the little ones were watching our activities through the windows from the dark side!

Finger prints on the left side of the car facing away from camp the previous night

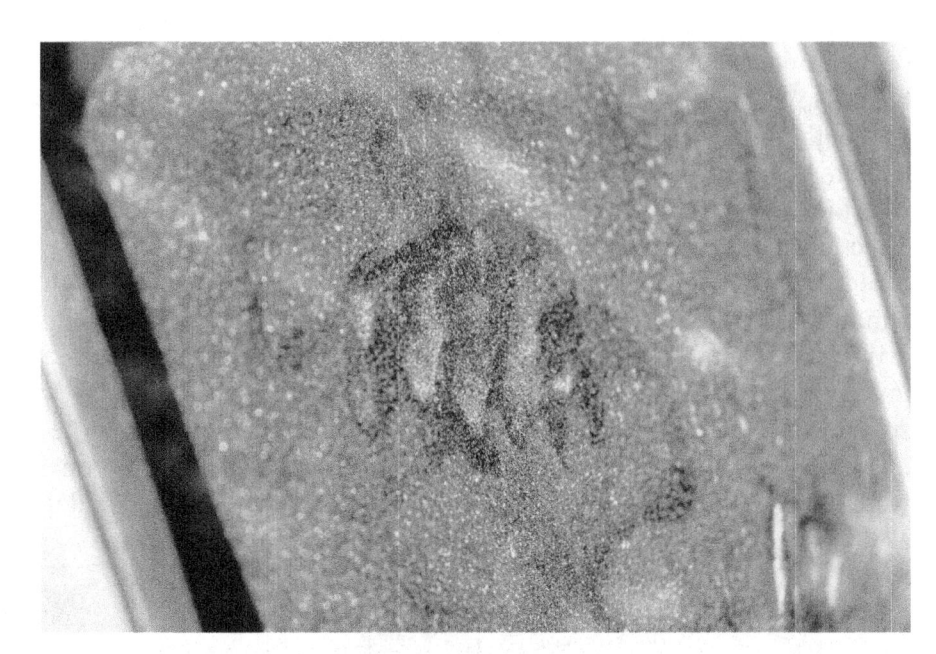

Closeup shots

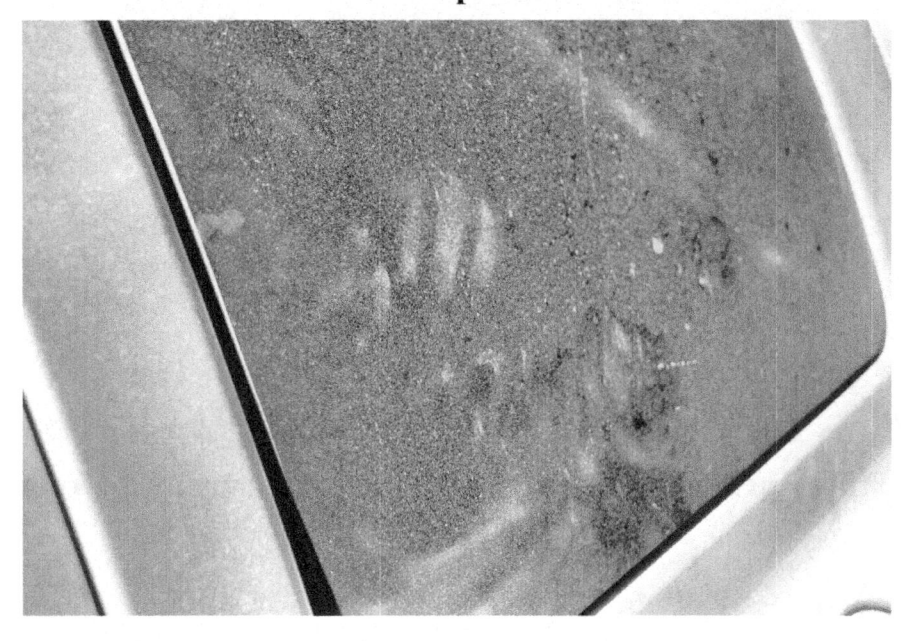

Later that day, Barb took a walk up the road above the spot where we were experiencing the eye shine and sticks/rocks being thrown at us the previous night. She and few of the other campers came upon a giant log structure in a teepee configuration (see below). There were four legs to it making it virtually impossible for this to have happened naturally. According to one of the campers Robin, who had seen these in Colorado, felt that it meant, "This is my spot, my home!" Also, you can get a pretty good idea of how large the Sasquatch is based on this structure's size compared with Robin standing next to it on the left!

Robin Roberts, a field investigator for Sasquatch Investigations of the Rockies (*SIR*) standing next to the giant teepee structure

Throughout that day, there were signs that they were all around us. Knocks were heard and a large chuck of wood was thrown from the brush. At one point, another camper saw what she described as an elderly face along with a youngster, in a tree just a few feet from camp. After dark, Barb joined three others for a walk into the forest above camp. One of them started playing the flute and called out to the Sasquatch using a language they may have been familiar with. He stopped playing and said, "I see movement in the trees!"

Barb agreed she saw it too. They tried to invite the Sasquatch over, letting them know they weren't afraid of them. The one playing the flute went ahead and introduced himself and the others since it was obvious they were right there though not seen.

Based on the movement, he felt there must have been at least twenty of them as it was hinted that the clan there was a large one. All the blinking eyeshine from the Sasquatches looked like Christmas lights in the trees - that's how many there were! At one point, they noticed something crossing the path in front of them.

Afterwards, Barb went back down the hill to join a planned night sit at an unused campsite. She first stopped by our camp to grab her folding chair and walked down to the designated spot where the others should have been. No one was there but she could hear voices further down from where we found the giant teepee structure. Having just come down from the hike on the hill, she was too tired to walk back and, because her legs were still sore, she decided to go ahead and set up her chair next to the fire pit and sit there alone in the dark. Barb's thinking was if there were any vocalizations that were being made; she would be in a good spot to record them away the rest of the group. As she sat there in silence, she noticed some eye shine in the nearby trees and up the hill closer to the others.

The main road was situated above the spot where Barb was sitting and, after a few minutes, she heard a car start and move past her towards our camp. The folks in the car were part of our group that had driven further to check out the area a couple of miles up the ridge.

After that, everything got real quiet as the eye shine increased with Barb seeing both red and gold colors beaming through the trees. Suddenly, something ran very fast across the tree line below her and, after a few seconds, she started getting a tingly feeling on her left side. It was something she never experienced before, as her whole left side felt just like every hair was standing up! As she looked back towards the right side of the spur road some thirty or forty feet away, she saw what looked like a brownish shadow that appeared to be upright and rocking back and forth. Although it was dark, moonless night, she could still see the shoulders lit up by the star light. At first, Barb questioned herself if what she was seeing was really there or it was just some shadows playing tricks on her.

Trying to avoid staring, she continued to sit there looking in every other direction including to the stars. Every few seconds, she would glance back and sure enough, each time she did, it continued to stand there rocking back and forth. While this is happening, her whole left side continued to be lit up with tingles. This went on for several minutes before things got really quiet again and whatever was standing there was now gone. [Note: The next day Barb returned to this spot with the others and found where something had run through the weeds as well as some freshly scuffed dirt and a fourteen inch footprint.]

Remember the car Barb saw passing by? Having remained in camp, I was there to see it as it was approaching our camp with the headlights on. I spotted what looked like a dozen red LED lights scrambling into the woods. Obviously, these were some juveniles that were hanging out next to a gifting spot one of our campers had set up.

After everyone returned to camp for the night, there was still plenty of activity during the wee hours past midnight with things like pine cones being thrown at the tents. There was the sound of running through the camp with one camper hearing what can best be described as a sucking noise, as if through a straw, and intelligent voices that sounded like Samurai chatter. I was also hearing something rattling the pans on my cook stove and footsteps walking back and forth past my tent.

On Wednesday morning, a fist print was found next to where some snacks had been left in a bowl on the gifting stump at the edge of camp. The snacks were now gone and the bowl was found below and to the side of the stump.

Later that afternoon, Barb, along with some others, went exploring in the forest above camp where they found a number of stick structures involving snapped trees. One small tree was completely snapped off and placed leaning against another, causing it to bow. Three dried mushrooms were also found on top of the remaining snap. The evening was spent at our camp sitting around in a circle playing Indian flutes and drums.

A short time later, one of our campers heard something in the bushes near her tent. She used a wood knocker to whack a tree twice to see if she could get a response. The group was in the process of walking down to the other campsite when a huge knock was heard at the edge of camp. She got her response!

We returned to the campsite where Barb had her experience the previous night. Since there were enough of us, we decided to take our chairs and form a circle with each of us facing outward. This way, we were able to watch in all directions if anything was going to happen. However, it didn't take long for the show to start. The Sasquatch were all around us providing us with plenty of eye shine as different ones caught sight of them. We also had a great time just sitting there watching the beautiful night sky with some shooting stars adding to our awe of its magnificence!

On Thursday morning, we noticed that the small gifting spot of goodies in a bowl was hit again but some marsh-mallows on a stick were untouched. Also found were a couple of red vine licorice sticks that looked like one of them had been sucked on. Perhaps the sucking noise that was heard the other night was a young one trying to suck the licorice like a straw. For the most part, Thursday was relatively peaceful. A few whoops and knocks were heard but nothing big or close to camp.

Thursday night, the group went for a walk. Not much was happening until they visited their favorite little unused campsite. They spotted several dark bodily forms moving about and at one point Robin's dog got spooked by one of them. There was also the presence of a skunky odor – a sure sign that the Sasquatch were near!

Later that night, I opted to go relax in my tent before trying to go to sleep. Though I was alone in this part of camp, I could still faintly hear the other campers hanging around the fire pit up a ways from me. However, I was also hearing some other noises close by that couldn't have been anyone in our group. Something was also moving around very close to the back of my tent. While I lay there, something was grabbing at my tent near the bottom and pulling it back and forth several times. I also heard numerous footsteps all around my tent.

Now my tent had a little bit of light still coming from the cabin porch that was making its way through the thin material. While I was looking in the direction of the light, I saw a strange silhouette of a head and shoulder moving very quickly in front of me. It had a very well defined outline though a little shorter than myself and looked as if someone quickly pulled a cardboard cutout in front of my tent. The weird part of this was there was no "up and down" body movement like when a person is walking or running. The figure moved horizontally with absolutely no vertical movement!

After sitting around our camp's fire pit, Barb also decided to call it a night, but first, walked down to the little gifting spot to check for any activity. She noticed that it had already been hit by the younger Sasquatches and then passed by my tent to let me know about it before heading to her own. It took no longer than ten minutes for her to get settled in her sleeping bag and shut off the light. Within seconds, Barb heard this huge thud between our tents and, at first, assumed that something had fallen from one of the trees overhead. Another few seconds later, Barb heard this high pitched, little voice calling out "Me-Me-Me-Me-Me" numerous times just a few feet outside her tent. This lasted around twenty seconds and after that, she could hear continuous footsteps all around her tent and back and forth between our two tents.

A few minutes later, she heard a door open as someone came out of the main cabin a few hundred feet up from our tents. Whatever was standing next to Barb's tent suddenly stopped moving while the person from the cabin made their way to the outdoor rest room. The second the person went back into the cabin, the activity just outside of Barb's tent resumed. The movement outside the tent lasted at least twenty minutes while Barb laid on her back trying to see if a shadow from one of them would show up on the tent but there wasn't enough light to see anything.

The next morning (Friday), we found numerous rocks (three of them in a row) and a piece of brick between our tents that weren't there the previous night.

Friday evening was great fun while we all sat around the fire with entertainment provided by several of our campers including yours truly on the guitar singing a John Denver tune.

After dinner and the entertaining music, Barb and some others picked up their lawn chairs and headed back down for a night sit at the little campsite where a lot of activity had previously been experienced. Once there, Barb noticed I wasn't with the group and headed back up to our camp to find me. I was actually in the main cabin having a great conversation with Thom Cantrall when Barb walked in to tell me about the night sit and invited me to come on down when I was ready. As Barb started walking back through our camp in the pitch black night, she glanced toward her tent and saw two glowing red eyes at least ten feet up from the ground staring at her! Barb continued walking while momentarily maintaining eye contact with the Sasquatch's glowing eyes thinking, yes, they're waiting for me! As she looked away and glanced back, the eye shine was gone so she continued on to the night sit.

I finally made it to the little camp and found that everyone had decided once again to position their chairs in a circle with each person facing outwards, leaving room for Robin's dog, Roo, to be comfortably situated in the middle. It was a fairly uneventful night although several caught glimpses of eye shine and movement in the shadows here and there. However, it was a beautiful night to do star gazing, watching for shooting stars and unusual light movements in the sky.

By Saturday, most of the group had packed up and left with but eight of us still remaining. It was sad to see people we had gotten to know over the course of the week leave one by one as the day progressed. Once everybody was gone, Barb took a little walk with Robin and her husband, John, and returned for the afternoon to relax in their chairs under the shade of the trees near our tents, reminiscing about all the experiences we had during that week. While they were sitting there, Barb heard what sounded like Japanese talking coming from the other side of some brush bordering a creek on the edge of camp. Barb remembered saying, "Well this is a strange place to have

Japanese tourists out here in the bushes!" ...Especially considering that there were no other vehicles parked nearby and there was no one around other than from our immediate group. It was only later when Robin heard the same thing on Saturday night that they figured out it was the so called Samurai chatter (language) of the Sasquatch!

Late Saturday afternoon, I drove up to the ridge overlooking the valley below and took some video and photos of the tree line as it descended down into a grassy field toward the valley. [After returning home, I discovered that one of the adult Sasquatch was looking out from the trees as will be seen at the end of this chapter.]

Earlier that day, I had taken my tent down, having had enough of the cold nights, in order to stay in the main cabin. Barb decided to stay in her tent for the last night, just in case there was going to be any more activity from the Sasquatch. After dinner, most of the campers including myself decided to call it an early night with most being in the cabin and Robin in her tent next to the creek. The only ones still up were Barb and Robin's husband John as they sat around the fire pit, waiting for one last, large piece of wood to burn itself out. They continued chatting over the barely burning log until at least around midnight. John was facing toward the forested hill while Barb was facing in the opposite direction towards her tent.

Suddenly, they heard what sounded like the loudest stomping Barb ever heard in her life going from left to right down by the bottom of camp near her tent! The stomping was loud and it was fast, the impact was so hard on the ground that both of them could feel the vibration with each stomp all the way up to the fire pit! Barb, since she was facing in that direction, could see a massive black shadow moving across the camp. John also turned to look that way and then looked at Barb with the widest eyes of amazement saying, "Now THAT was a Bigfoot!"

Getting very excited about this, John hurriedly poked at the fire with a stick, trying to get it flamed up again. Barb then said to him, "Well you know, you realize now we can't see any farther than beyond this ring of light."

Immediately, John does an about face and starts scattering the wood around to put the fire out. The fire pit was now down to glowing coals. Five minutes passed and the stomping started back up again but this time it came from the right of camp and made its way to the left.

Another ten or fifteen minutes went by after the second stomping incident before Barb decided it was time to go to bed herself. She left John by the fire pit which at this point had pretty much burned itself out. She walked up to the restroom next to the main cabin. As she left and was walking next to the cabin, she heard the loudest "Whoop" sound coming from below the camp by the creek near the gifting spot. It was so loud that it echoed through the entire camp! Barb was now thinking, "Holy Smokes! What's going on? Why is John down there doing whoops in the middle of the night while everybody's sleeping?"

As Barb took a couple of more steps past the cabin she could see the fire pit and John was still sitting there. It finally dawned on her that it wasn't John after all, it was the real deal! Barb walked back over to John, being stoked by the events of the night, and helped him pack some of their stuff back to their tent. We found out later that Robin was still awake and listening to all this activity. She later sent us her account of what she was hearing while all this was taking place:

"I heard a conversation that sounded like Japanese behind my tent across from the little creek there while you and John were at the campfire. I also heard knocks several times. At one point something walked by the tent and Roo growled a little and followed it as it went over to where you and Sam's tents were. When you walked over to your tent with your flashlight I saw something upright run by my tent back into the trees. Then, just before Suzy and Diane came out of the cabin when everyone was in their tents and cabin, I heard a whoop and then a rhythmic knocking was starting until Suzy came out and was talking to you [Barb]. I was glad to go to sleep before everyone else but did not really sleep much. LOL"

Later, Barb returned to her tent and, being open to the total experience of that night, opened up every window flap. This way, if anything were to happen outside, she could look out to see them and they could, of course, see her as she was hoping for some kind of interaction. She then got settled into her tent and turned out the light. Her cot was positioned in such a way to allow her face to be only six inches away from the window screen. As Barb lay there, she continued to look outside with wide eyes, being a bit nervous from all the stomping and the loud whoop heard earlier plus the fact that I was no longer next to her tent. However, there were a couple of people from the cabin who were going to and coming from the outdoor restroom a number of times and we think this probably discouraged any real Sasquatch activity from that point on. Barb did hear the scurrying of the smaller young ones moving about camp a few times before she fell asleep.

On Sunday morning, we noticed the hand prints on my car window were showing up really well with the accumulation of dust and pollen over the past few days. The details, such as the dermal ridges, were very obvious and Barb was able to take some pretty decent photos for her YouTube channel.

I had a dilemma as to washing my car and losing all those wonderful prints. After I returned home, I immediately obtained some black construction paper and clear shipping tape. I was able to carefully remove the prints with the tape and placed them on the black paper before making my way to the car wash.

We finally left the Blue Mountains and headed back home, being very happy about all the experiences and the great friends we made while there. Personally, I was glad to get back to my regular routine and Gabby, who was not able to go with us, was very glad to see Barb again!

Here's an enlargement of a photo taken of the edge of the forest overlooking a grassy decline. Photo was taken on Saturday evening - our last night in the Blue Mountains.

Close up view of the Sasquatch. Notice the side profile with the human-like face, high cheek bone and white beard.

Chapter 14
Connecting with the Sasquatch

One of the things that was happening to me was the overwhelming feelings I was having as I would take walks in the forest. I knew I was empathic with some psychic abilities and I could literally *feel* the Sasquatches nearby and the sense of their desire to connect with me.

On one of my walks through the field of dreams on June 14th, 2015 I recorded a video of me talking as I walked through the forest in which different viewers on my YouTube channel later identified several Sasquatch standing next to the trail as I passed. Being in a meditative state of mind, the words seemed to flow as I began to speak. I transcribed the following paragraphs from my video word for word. In the first part of this I'm stating my own feelings about my personal research and people's expectations of the evidence I present:

"I don't expect a visualization or visual encounter. As any good Sasquatch researcher knows - you're not going to get a clear picture. I don't care how hard you try... they, the Sasquatch, got one up on us. We, as humans, move like turtles conpared to them and whether people believe it or not, they can distort electronic equipment. I've been fortunate enough that they have allowed me to take the pictures I have taken which, of course, they usually only appear in a very small spot on my photos. Not really hiding but just kind of there and it takes time to go ahead and find them sometimes."

"Of course I know what I'm looking at. And when they *do* show up in the pictures, you know it's not some abnormality of the photo. It's not a blobsquatch. It's got enough clarity for you to know - it's a Squatch! Just like the large Squatch (in episode 6) and the other one was the one relaxing by the creek. I mean, those were very obvious! Nobody's going to dispute that! I see people on Facebook or Youtube that will make comments like 'Anybody can make it out to be anything they like'. I, of course, disagree with that kind of reasoning. I let the viewers decide what they are seeing and let them be the judge instead of these self-appointed Bigfoot experts who are hell-bent on discrediting genuine evidence. The majority of people speak out and based on their comments and likes they are the ones who are overwhelmingly agreeing about what I'm presenting. People with common sense know better. Let the people be the judge."

"I do miss the frequent vocalizations we were hearing in the 'field of dreams' - it's just a quiet time of the year. It's the same thing last year when the activity seemed to be around the August and September timeframe (this once again proves to be true with some incredible experiences just a couple of months later!) So it's probably just the season. We still look for the other evidence like the wood structures and glyphs. Barb is still busy with her gifting spots and is having success with that. I personally have done gifting a few times but I don't make a regular habit of it. And, of course, looking for footprints, etc."

At this point, I began letting go and a feeling set in to help guide my words. I truly believe the feeling was coming from the Sasquatch themselves as they had some influence as to what I was saying:

"I just like being out here. I'm into it, not for trying to prove anything to anybody. I'm out here to ultimately get full interaction with them. There's a lot they can teach us. There's a lot of unanswered questions. Just the mere fact I know they exist can affect a lot of our current belief systems that are taught from birth on. It totally negates them just on the basis there are other intelligent hominiods on our planet. This is just the beginning... and I'm looking for those answers! I'm trying to gain their trust so that we can open a door. If enough of us gain their trust, they won't be so afraid to come to a certain select group of people. Maybe they're waiting for the right time. We know what the world is like right now, we know there's a lot of devious humans out there that will still cause the Sasquatch a lot of trouble if they try to come out to the general public. So maybe it's just a waiting game for the right time. In the meantime, the few of us that are trying, we want to show them we mean them no harm and that we care for our environment. I think that's something all of us need to focus on. It's hard to put into words but a lot of what I've learned since I've just been out here getting closer to nature is that there's a bigger picture here. Regardless of how we live our lives in this planet there's a lot of things that are not compatible with the sustainability of earth. We, as humans, are going to have to get used to the fact that our lives may have to change radically to be able to continue to exist! This is where I think the Sasquatch can actually teach us *how* we need to live or at least compromise and begin to head in that direction that's compatible for us to continue to exist on this planet. Otherwise, if we don't, no one is going to be living here."

"In a worse case scenario, the Sasquatch and other creatures who live in harmony with the planet will continue to exist here and the human race will simply die out. That, of course, is not a pretty picture so I'd really rather not entertain that prospect. Everyone needs to do their part in trying to *feel* what's the right thing, to feel nature, to feel harmony living with the planet and then to teach others. Get them to realize it. Bring them out, let them appreciate it. Let them learn appreciation for what they have here and to get away from the things that are causing the kind of destruction we see today. I feel this very strongly in my heart! I honestly believe they know that coming from me."

"There's a symbiotic relationship I think you form with them and I think a lot of times just the expression of thoughts are sometimes, I know they're coming from me but sometimes, I feel a connection with them. And sometimes, I feel that their thoughts are merging with mine and then I speak. But I think this is how they feel... seriously!"

Chapter 15
Apparition or Sasquatch Mind Trick

During a small campout we had at Barb's place in early August, something very strange took place. Although I had been there at her place earlier that day, I had to head back to my cabin to take care of some things for a little while. Afterwards, I left and went to the tavern for a quick beer.

The next thing I know, Barb walks in the tavern and says to me, "There are two things that I need from you: the coffee you have in your car and to check to see if you're OK..."

I looked at her kind of puzzled and said, "Why would you check to see if I'm OK?"

Barb replied, "Well according to Darin (one of the campers), you were there at my cabin just a few minutes ago!"

I told Barb that's not possible as I wasn't there for quite a while due to taking care of things.

"You need to come down and talk to him. He is convinced he saw you!" she insisted.

I left the tavern quickly to see what in the world was going on! Do I now have a twin or doppelganger hanging around Greenwater? It only took a minute to get to Barb's place and as I got out of my car, I saw Darin sitting at the picnic table. I immediately noticed he looked a bit disoriented or maybe even in a subtle state of shock. We both walked over to his tent on the far side of the property he then related to me what happened:

"So I was laying in my tent and happened to kind of just look out, not getting up, and saw... YOU!... at the picnic table. So what I was thinking was, while you were outside, I thought everyone was talking and I saw [don't remember if I saw or thought I saw] Barb and Sandy but I definitely saw you. So I didn't pay any attention as I was thinking 'OK... time to get up'. So I sat up, got my shoes on, opened up the tent, wasn't looking up there, got out, and started walking to about... here I would guess..."

Darin was now standing a few feet from his tent in direct sight of the cabin and the back window with the picnic table between us and the cabin.

> "And I noticed... *you*... and in both times, it was your back and it was kind of a grayish charcoal shirt (I *knew* you had the Sasquatch cave art type of shirt. I have one too, so I knew that shirt!) So about here [we are now ten feet away from his tent] is where I saw you in the window!"

"I thought you were *in* the house when I saw you in the window so then I walked up to the picnic table but it was getting dusk, darker so I was paying attention to the ground where I was walking. I didn't pay any attention to you guys so I walked right up to the picnic table and noticed that it was all cleaned off... there was barely anything on it. There was my water bottle and one or two things on the other side of the table because we already had dinner and I think everyone already cleaned it off and because there was nothing on the table, I assumed you were in the house and people were doing dishes... Sandy and Barb... or you were just talking. But when I got to the table, I didn't see anyone or hear anyone! It was just now turning dark so I opened up my water bottle, drank some water, and I started thinking well.. OK! I don't hear anyone in there. The table's clear and everything seems to be in order so I thought maybe it was just Samantha [referring to me]. I happened to go up on the porch, opened the door, stuck my head in and noticed that the bathroom door was wide open [being visible from across the top of the refrigerator on the other side]. I said 'Hello?' (since I saw you in the window and just assumed you were using the restroom with the door open thinking that no one was around). No one said anything so I closed the door and came back and right when I turned around, I noticed a red flash and I heard Gabby over by the cars so I walked down and had some more water."

"And that's when I saw Barb and Sandy talking over there and I sat down and then they came up. At that point your car wasn't there and you were already gone, you weren't even around!"

I said, "I wasn't even here! It's kind of weird because I had already left earlier to go back to my cabin and take care of some things." I then related to Darin how Barb came up to the tavern after spotting my car and asked me if I was OK.

I continued, "That is so bizarre that you saw me not just once but twice... First standing in front of the picnic table and then, just inside the house where I can be seen through the back window!"

Darin repeated what he saw with more emphasis, "I was lying down trying to take a nap but was definitely not asleep! While I was still laying in my tent back there, I just happened to look without sitting up... just looked out my tent because I could see all the way up here [the picnic table] and I saw you standing right here at the end of this table! At that time I just remembered seeing you. I don't know or I assumed you were talking with Barb and Sandy."

I asked Darin, "You didn't see anybody else? You were just

seeing me standing there?" Darin states that he only saw me at a distance by the picnic table while he was looking out the tent. Once he was about ten feet away from his tent and closer to the cabin, he then spotted me in the back window.

Darin made a point concerning the cabins on either side of Barb's place from his view just outside his tent. On the left side, there were people talking and laughing in their back window and on the right people were sitting on their picnic table next to the dry creek only fifty feet away. So Darin was surrounded by people while this happened.

Was it a Sasquatch that came to visit and made Darin *think* it was me? Was this an apparition or mind trick of some sort related to the Sasquatch or some of the other strange activity we experienced nearby? Though I have my own opinion on this one, I'll let the reader make their own determination of what this could be.

Chapter 16
A Personal Message from the Sasquatch

So while we were still camping at Barb's place that Friday night, there just so happened to be some activity a quarter mile away at my cabin which involved George, the individual that was staying in my RV. I found the next morning that George was abruptly awakened from a sound sleep around 3am Saturday morning. He had the window of the RV cracked open and as he explains it, he heard whooping howls. After a few howls, he got up and walked to the door and started yelling, "Sam! Sam! Are you out there? What are you doing? Are you playing a joke on me?" Now George knows my involvement with Bigfoot research and that I had, on occasion, done a few whoop howls to show him how the Sasquatch sound. Of course, George is steadfast in his disbelief of Bigfoot so at the time he heard these loud howls he just naturally thought I was out there pulling his leg.

Waiting for me to respond back, he heard three more whoop sounds coming from the strip of woods right behind my cabin. What's interesting about the woods around my property line, it's only approximately twenty to twenty five feet wide and completely spans the perimeter of the backyard. There are houses on the other side of this strip except for a narrow piece that goes up to the main highway and then crosses into a very forested area. Obviously, this is the pathway the Sasquatch take to enter my property and around as was the case of them watching my neighbor's TV. At the entrance of a small trail behind my cabin, I had recently created a standing "X" structure as my "Welcome Mat" for the Sasquatch that drop by for a visit.

It was about this spot next to my X structure is where George had heard the howls:

I tried to reassure him that it was not me doing the howls nor would anybody else be making these kind of sounds at that time of morning. Of course he's still convinced it had to be some kids partying into the wee hours of the night so I didn't argue with him. I left the cabin to go back to Barb's until late evening but returned to my cabin right before dark. As I walked over to my Bar-B-Que grill and looked down at what I remembered to be a random pile of sticks, they had now been transformed into discernable glyphs! It's a good thing I had my camera to take a few pictures of this as I was a little upset the next day when I discovered that George had for some reason raked up the sticks.

Shown below was the entire glyph sentence made of sticks:

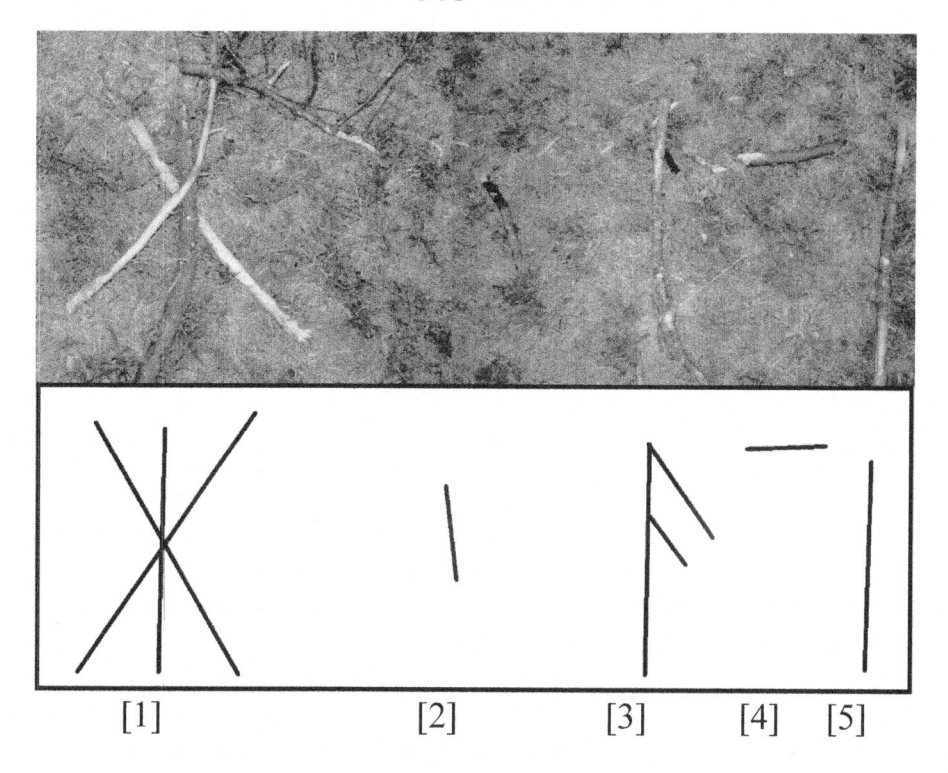

[1]   [2]   [3]   [4]   [5]

This was definitely a message to me due to the inclusion of my personalized glyph [3] (the Ansuz symbol). The Sasquatch *knew* I was across the way but took the opportunity to give George a personal experience and to leave me their calling card by way of this message!

Chapter 17
Morphing Bigfoot Incident

Almost a year had passed since I witnessed the light portal and orb and I had yet to publicly report the incident on my YouTube channel. I hesitated due to the fact that I had very little proof other than my word and the photo of the sheared triangle on the ground. I did, however, disclose the incident during a presentation for the Team Squatchin' USA Bigfoot group headed by Dr. Matthew Johnson a few months after he, along with several others and at different times, had witnessed an amazing portal at a location referred to as SOHA in southern Oregon.

As time passed in my research, it became clear that any research involving the Sasquatch would also lead to unexplainable events that could not fit neatly in the "scientific box" that many old school researchers would confine themselves to. Anything that fell outside of this "box" would either be ignored or dismissed as invalid evidence. Also, during my earlier involvement with the general Bigfoot community, I felt the peer pressure to conform to preset limits on what you could or could not report. Anything that was unusual or "out there" would be met with ridicule by certain individuals who felt they had to police or censor information, pictures or videos that did not help to establish that Bigfoot was a real animal running around in the woods and certainly a lessor being than us humans. The threat to them was to suggest otherwise, that the Bigfoot were, in fact, a race of intelligent human-like beings with their own sense of community and spirituality.

Thanks to people like Dr. Johnson and others who have not held back on reporting the strange and unusual experiences, often referred to as the "woo", I decided it was time to come forward publicly on my experience of the light portal and orb on video. I felt it would be best to tell the story while I walked and recorded video along the same path in the field of dreams as it happened.

On August 27th, 2015, I headed to the area and stopped at a spot on the trail where I would start recording. I was amazed though at all the crazy activity up on the ridge with stick knocks while I set up my camera. I remember saying out loud, "I know! I know! I know you guys are up there but I'm here for a different reason and I need to get this done!" So I passed up the opportunity to focus on all that activity just so I could tell my story about the portal! I proceeded to record the video as I walked slowly toward the spot where it appeared and, finishing up, I headed back to my cabin to edit the video. I added a simulation of what I saw the prior year and posted it to my YouTube channel as "Episode 35 – Light Portal/Orb Encounter!"

A few days later, a regular viewer of my videos made the following comment:

"Am I going crazy at 6:35 minutes in, or is there a creature almost in the center screen between the two larger trees showing off its upper body for thirty seconds or so? Blends in very well but I see movement, cone head, clear eye brows, face and upper arms from something?"

I reviewed the video slowly and was shocked to see this large head pop up just like what the viewer reported. Although the camera was moving, it was still pointed in the same general area the creature was and I was able to observe it for a half a minute before it moved behind the tree.

I undertook the task of creating a series of screen shots showing not only the different movements being made but also changing facial expressions. As I examined each shot carefully, I also noticed something else – the head seemed to be morphing into different shapes as well! On the following page are nine of those screen shots and starting with the first one on the left to the right and down, you will easily see the head moving but also changing shape. On my video "Episode 36 – Amazing Sasquatch Activity on Video – Part 1", you can watch this movement in slow motion.

From my personal observation, it starts out looking like a Grey Alien with large eyes before changing into something that more resembles a Sasquatch. Was this a Sasquatch or a Grey? I couldn't tell you for sure but perhaps the Sasquatch on the ridge were really trying to warn me that something was in the field of dreams before I started recording! Although this left me with more questions than answers, it would not be the end of the story as you will later find out.

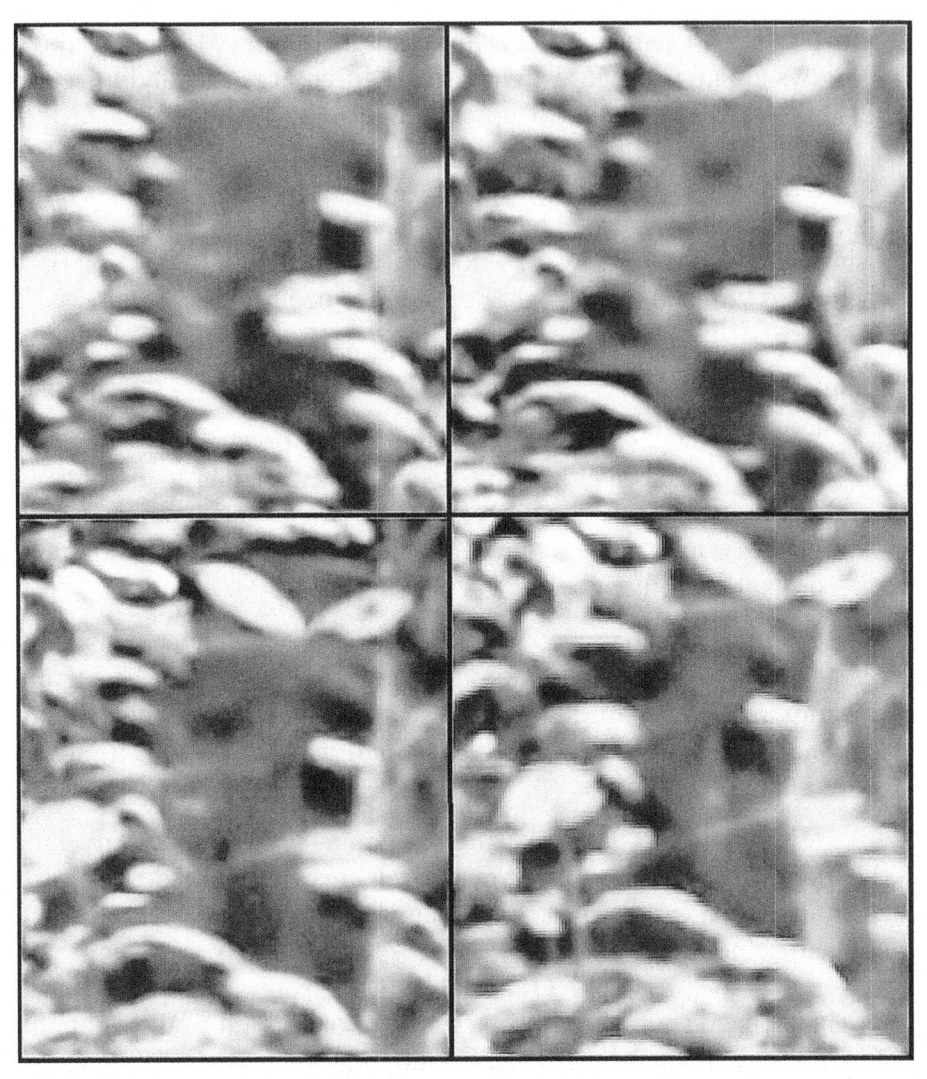

Several screen shots showing how the head changes and turns as I approach it while taking video. A high definition color video is available on the Planet Sasquatch Youtube channel where the movement can be directly observed.

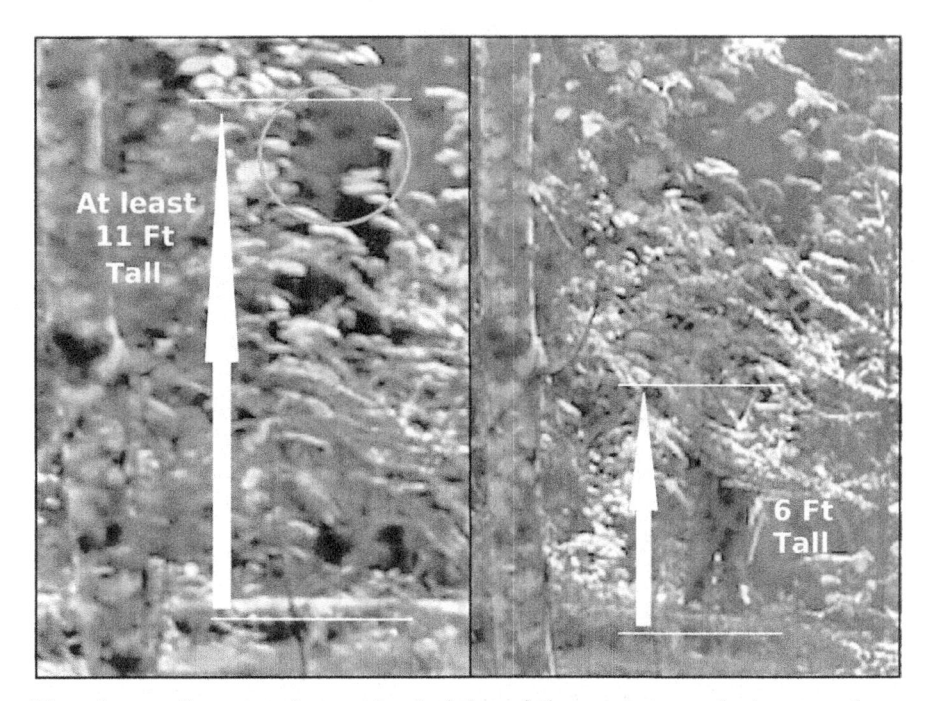

The above diagram shows the height of the creature relative to a later photo of me standing next to the same spot. Although I stand at nearly six feet tall, the creature easily towers over me at almost double my height at eleven, possibly even closer to twelve feet.

Additional Close up Shots

Chapter 18
Sasquatch and the UFO Incident

During the Barb and Gabby Campout in early September of 2015, there was a couple plus the woman's daughter who were camping across the way from us. Being friendly, they came over and struck up a conversation with Barb and the others with us and were commenting on the big bear they kept seeing occasionally. They had actually been there for over a month and had also heard strange howling sounds especially at night. After giving us a complete description of what was seen and heard, we came to the conclusion that what he and the family were experiencing was really a local Sasquatch clan. John, the male member of this trio of campers, at first scoffed at the idea. However, when we explained that we were there for the expressed purpose of doing Bigfoot research, he felt it was a sign since he was getting perplexed trying to figure out what in the world this creature really was. Telling us initially it was a bear was the only thing he could think of from his standpoint.

Starting the very next night, the activity around John's camp started picking up during the late night, only for us to hear about it the next morning as he came over to help chop wood and recount what they seen and heard. It's almost like the Sasquatch could read John's mind. They probably felt that since John now knew who he was dealing with and that the Sasquatch would be no threat to his family then it was time to start doing some serious interactions with them. Both John and his girlfriend recounted how, while sitting around their campfire, they would hear them coming in close to their camp until, as they looked up, they could see several small juveniles sitting on a dirt mound within fifteen feet away... just staring at them! To the right of their camp, there was a steep hillside and due to the light coming from the fire, they could look up and see several members of the clan sitting on the hill. They arranged themselves in rows segregated by status. The larger Sasquatch, presumably the adults, were at the top. The next row is where the older juveniles were found while the bottom row... the one closest to the campers... were the young small ones.

However something strange was going on with John as each day passed. His growing knowledge about the Sasquatch did not sound like a man who had just found out that Bigfoot was the real deal just the other day. Based on what he was saying was very similar to the knowledge I had picked up in my own interactions with the Sasquatch over a year's time. John learned things about them in a fraction of the time it took me! It was important for me to find out more through deeper conversations with him, to try and figure out what was going on.

"They get into my head" he said referring to the movie clips that were projected into his mind. The Sasquatch were conveying knowledge to him via these clips and at times, John heard the clan leader speaking to him telepathically.

At some point while he was sitting at his camp, the larger Sasquatch came by and touched his hand. Later, he had the sensation of a warming hand on his back. According to him, the touch took away the acute pain he had been suffering with for years and was now able to go about his business without pain medication. On another occasion, he felt like one just brushed by him leaving a mark on his chest for whatever reason.

Our groups's campout ended on Sunday after four days and we packed up and left. John and his family continued to stay in their spot just as another couple set up camp across the road. I was still intrigued by their continuing experiences and frequently stopped by in the evening right before dark and hung out by their campfire.

About a week later, I ran into John's girlfriend in town and she asked if I was coming out to the camp that night. "You just have to come over tonight! John has something he wants to show you... it's unbelievable!" she insisted.

"What is it? Are the Sasquatch coming into the camp directly interacting with you?" ...wishful thinking on my part, but I can hope.

"I can't tell you... you have to see it for yourself!" as the excitement in her voice picked up.

By now, my curiosity was going on overdrive! How could I pass up a once in a lifetime chance to see something that I had no idea of what I would be shocked by. Her insistance on me coming over that night would guarantee that it would knock my socks off!

I later dropped by and sat by the fire. After some general conversation, John leaned back in his chair and closed his eyes silently meditating for a few seconds. Suddenly, we heard the loud calls of Barred Owl imitations typically used by the Sasquatch in this area and it was *very* close! After about fifteen seconds it abruptly ended.

John then opened his eyes and said, "Did you hear that? I asked them to do that!" I was not doubting this as it seemed to be planned between the Sasquatch and John ahead of time.

It was a starry, clear night and I asked him, "Was this what you wanted to show me? I am very impressed!"

John's girlfriend, who was also sitting there said, "Oh no... you will not believe what he is going to show you now. I witnessed this for myself several times last night!"

John got up and walked past the main tent to a spot where the sky was clear to see. I followed him until we stopped and he began to look around intently at the sky as if he was searching for something. After a short pause, his attention stopped to focus on a particular spot in the sky. He proceeded to point at a particular faint but visible star making sure that I had the right one myself to focus on.

"Keep your eyes fixed on that particular star and watch!" as John's stare got intense. As I watched the star, he then moved his hand while pointing with his finger slowing from right to left. In my utter amazement, the "star" now *moved* in the same direction following in sync with his finger. Talk about a jaw dropping experience! He then moved his finger in a "U" pattern and sure enough, that light in the sky also moved in the same pattern (notice that I am now convinced I am not looking at a star). What I saw gave me a flashback from my past as a kid holding an Etch-A-Sketch and using my fingers to control the two knobs to create a moving line on the screen. Finally, he whipped his finger quickly to the left and the light followed accordingly... disappearing from sight. I kept staring at the sky with a smile on my face grinning from ear to ear.

Who in the world (or perhaps *another*) was this person I'm standing next to that can move star-like lights in the sky as if they're waiting for commands from the captain of a ship? (Speaking of a ship - that is what I determined it to be... a UFO of some unknown origin.)

I then turned to John saying, "This is great! I would have easily paid you at least two hundred dollars to see this event!"

As we walked back to the tent, his girlfriend began to tell me that he was doing this at least seven times the night before. Another couple who was camping across from them had also witnessed this the previous night but their reaction to John's abilities did not set well with them, as they sank in their chairs, possibly faint out of fear of what was happening! For me, there was no fear about this event or even concerning John's *unusual* ability. It was just further confirmation of what I was already experiencing with the Sasquatch and my encounter with the light portal the year before and the shape-shifting creature from the previous month.

We finished the evening sitting next to the campfire as I was very intent on asking John many questions. It's not everyday you see humans point at the sky and control UFOs by the snap of a finger (no pun intended). John began to reveal more about himself and his past and what he was able to do as a child:

> "I was given up by my mother to live in a children's home at an early age because she could not handle my special gift." I asked what special gift he was referring to. "I could move things like pencils from across the table without physically touching them and she could not handle that. That's the reason she put me in a home though that's not the reason she gave for doing so."

Hearing this, I had no basis for doubting him for I had just witnessed one of the most remarkable things I had ever seen in my life! I replied, "Telekinesis, the ability to move objects with just your mind. It makes perfect sense to me, I know this is for real!" I speculated that he had somehow temporarily gained control of the ET's ships by mental thought. I continued, "I don't know, John, I'm a little worried that someone up there is going to get a little upset with this and is going to come looking for you on the ground. Especially if they obviously have advanced technology, it wouldn't take them long to determine *where* you are."

Also, even though John was able to easily control the seven ships the night before, he was only able to do so with the one that he showed me. It occurred to us that "they", the ET's, were getting wise to him and taking measures to prevent the unauthorized access of their ships. Afterwards, things settled down and we had a few cups of coffee around the fire before I left.

Returning the next morning John told me he had something happen to him after I left the night before. Sometime during the early morning while still dark, he got hit with "something" that caused his body to feel on fire! His girlfriend said she literally saw "steam" coming off his body and that his eyes turned glowing red! This lasted for a couple of hours before subsiding.

I looked at John while shaking my head and said, "Hmm... So I guess they found you after all. You're still alive – that's a good thing! I guess this was just a warning shot for you to knock it off."

He agreed. There is no sense playing in something you know absolutely nothing about, especially Extraterrestrials with advanced technology.

As fate would have it that night, there were two lights, much like orbs or perhaps drones with piercing green and white lights, that were peering through the trees all night and were getting close to the campers while they were in their tents. Being alarmed, they retreated to their truck, started it up and the lights then seemed to back off. Once they shut the truck off, the lights would again get closer until they started it up again. This went on all early morning long until it was light. They were completely exhausted from the experience when I arrived in the morning and they told me what happened.

I wanted to come by one last time that following night as it was the blood moon event and, from their camp, I could get a pretty good view of it. Since the next one would not take place until the year 2033, I did not want to miss it. After taking photo shots of it as it was waning, John looked over in the trees and said quietly, "They're back!" He then directed my attention to deep in the forest across from their camp and I could see, though faintly, the green and white lights he had seen the night before. They were in fact moving back and forth slowly but never came closer while I was there. Before long, the lights did disappear and did not return to bother them. We never did determine what they were and probably for the best.

Afterward, on another occasion, when I wasn't there, John and his girlfriend, along with another couple, walked to the bridge crossing a small creek after dark. It was still a full moon and you could see things fairly well up and down the creek. This was also the time when the salmon had returned to spawn so the creek was full of them. As John looked down to where a log crossed the creek, he spotted what at first thought was a bear standing up. As both he and his girlfriend stared at the creature, its body began to move back and forth while stepping very slowly toward the side of the creek. As the light of the moon began to reflect off of it, John and his girlfriend knew that this was not a bear but a large Sasquatch they were observing! The other couple were in disbelief and keep trying to insist it was just a log that was sticking up out of the water. They kept watching it for a good fifteen minutes before leaving. Later the next day, John walked back down to the bridge and sure enough the so-called log that the other couple claimed was in the creek was no longer there!

John and family eventually broke camp and left that area. However, there is one last thing worth mentioning while they were there. During the mornings, especially while they were sponge bathing at the back side of the tent though fully visible from the trees, they had a weird feeling of being watched. I had arrived after they had finished bathing and looked up into the trees on the back side. One of my abilities is to know when *someone* or something is there in the background. I knew by certain darker shades or shadows that there were possibly small Sasquatches watching from the tree. I grabbed my high resolution camera and took a couple of pictures of the background. Later, while on my computer enlarging the images, I found one being I assumed was a Sasquatch looking down at us. After minor enhancements to bring out some of the detail, though not as crisp and clear as I would have liked, I ended up with the picture of its head as shown on the next page.

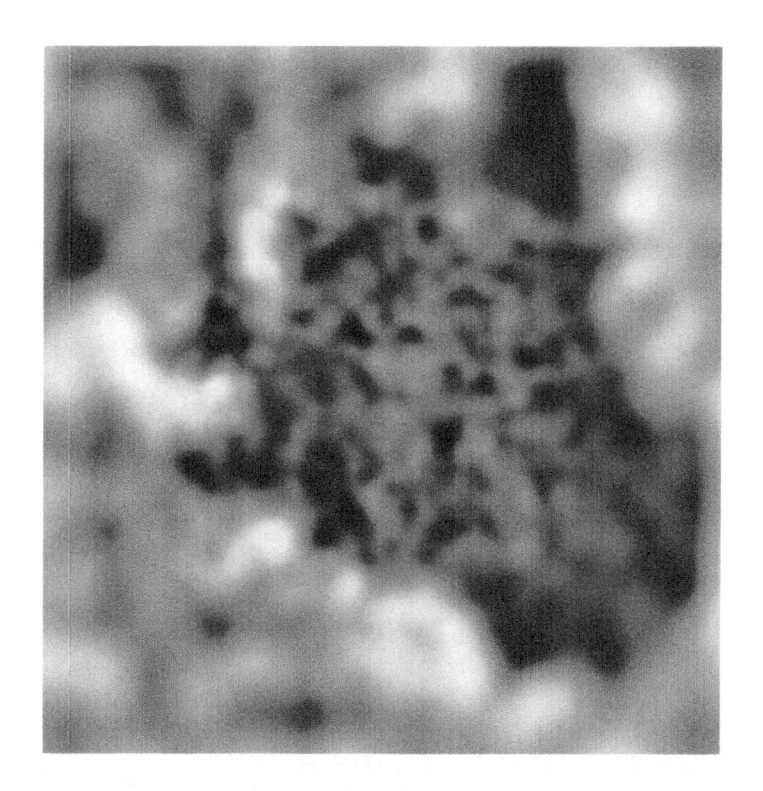

When I later showed John the image of this fellow, he said, "That's him! That's the being I'm seeing on the ships!"

I asked him what he meant by this and he went on to say, "Remember how I told you I could see movie clips in my head? Well, when I took control of those ships, I could see a movie clip of who was up there and they could see who I was!"

Maybe that's how they ended up finding John that other night in order to "chastise" him.

This was just one more piece of a puzzle that has left me mystified. This whole experience involving this family's experiences has generated more questions than answers. Who are the ET's of the ships that John temporarily took control of? Why do the Sasquatch seem to be connected to events like this? Why are certain humans like John exhibiting supernatural abilities while most don't? Only time will tell and I am determined to find some answers!

A sketch drawn by my daughter based on what she saw in the photo from the previous page

Chapter 19
Two Beings Appear Behind Me

There was something else that happened around the time the Barb and Gabby Campout was going on in September. I had made one of my usual trips into the field of dreams by myself. It was September 9th, 2015 and a little over two weeks after I had recorded a video while walking here and relating my experience of seeing the light portal and orb back in 2014. Once I posted the video on Youtube, a viewer pointed out what I later identified as a shape-shifting creature that at first looks like a Grey Alien and morphs into what appears to be a Sasquatch! This particular event was discussed back in chapter 17.

All this was fresh on my mind while I walked down the trail in that narrow valley surrounded by cottonwoods. With less than a hundred yards to walk through this clearing, I began to sense that I was being watched! I turned around and took a couple of photos of the little valley I just walked past and then started recording video of the surrounding area with my Nikon camera. I panned the area slowly as I was talking and was rambling a bit pretending to be taking video of the beautiful surroundings but in fact was purposely scanning the field of dreams for whatever was setting off my spidey senses!

As with many times before, I usually discover things after the fact and upon closer examination of one of the photos I took, it gave me the answer of *who* was watching! After enlarging the photo, I began to notice a dark figure standing to the left of the field. As I was trying to do some enhancements on it, I almost missed a second figure that was blended quite well with the background as you can see on the next page.

The only thing that I could make out on the right was this slender being with sloping shoulders and shrouded head. As I said, the one that was most apparent when I first looked at this photo was the black figure on the left. Looking closer to the right, you can barely make out the other being, although the black and white photo for this book makes it just that much harder to see.

Here are the two beings side by side without the surrounding background and enlarged:

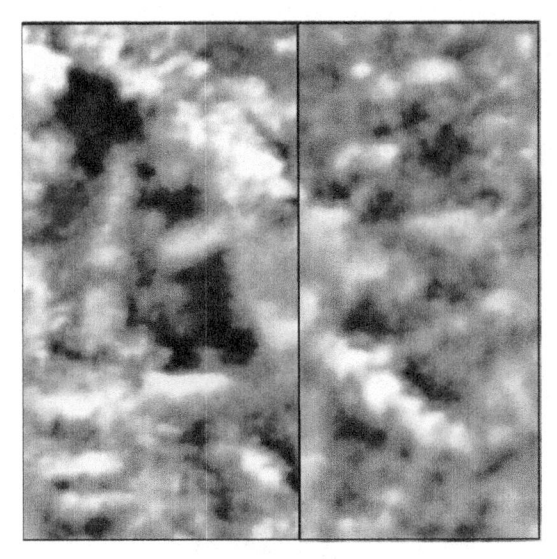

Even though the dark being on the left was easy to identify, I found it difficult to make out any distinct facial features other than seeing the head with a lighter area for the face and possibly the mouth. However it's a different story for the being on the right!

Below is an enlarged image of what I believe to be is a female Extraterrestrial being of some sort. Clothes are apparent with a wide collar around the neck. The eyes are very large with a narrow face and small chin. The ears appear to be pointed up and out and there also appears to be some kind of ornate wrap around her hair. I had my daughter do a sketch from this photo and it is very close to what can be discerned.

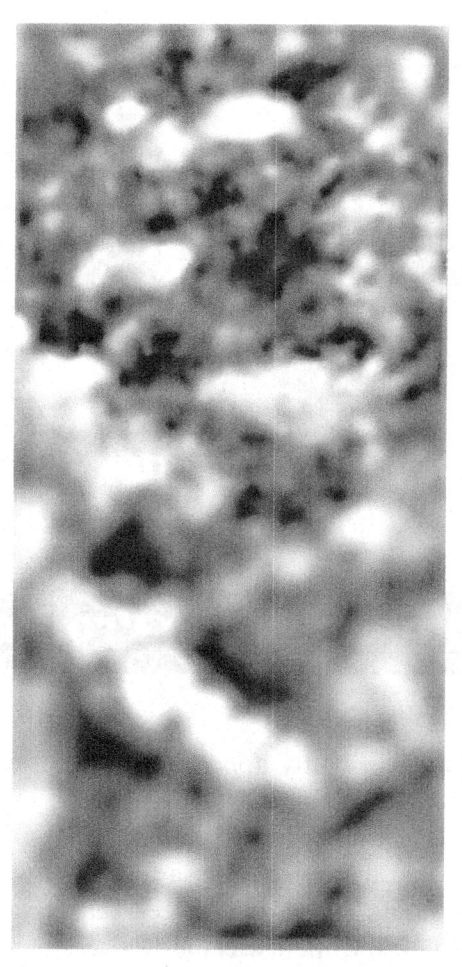

My daughter is a really good artist so I let her do a profile drawing of what she was able to see in the close up of the photo below. This is very close to what I'm seeing as well. We have a clothed being with a pleated wide collar top, thin neck, triangular shaped head with large eyes whereas the pupils are more slits than round, thin chin and mouth area, pugged nose, larger ears possibly pointed, very light to white hair with some kind of patterned cloth bandana wrapped around it. From the full body image on the previous page there is a smaller one who seems to be hugging the being possibly alarmed by the fact that I'm walking toward them (unbeknownst to me) and taking pictures.

Photo Comparison to Profile Drawing

How did they instantly appear in that spot after I just walked past it a minute earlier? They also disappeared just as quickly as I walked back while taking video! Based on comparison photos I was able to determine that they were only a few feet away from the portal spot. Did the morphing entity some weeks earlier also use the portal to come and go quickly while I was there?

I have more questions than answers on this one. Where are they coming from... Another world? Another dimension? ...Or perhaps another time? Is there a connection to me and was I *supposed* to see all of this? I feel these answers will come eventually as I strongly feel it is related to the intuitive direction from the Sasquatch. It is apparent to me that the Sasquatch are working together with other beings that we would consider Alien or Extraterrestrial. When the time is right, I feel that we will also work with them in a more direct way!

Chapter 20
The Habituation Method

It's interesting to note that the term "habituation" is defined as a process where there is a decrease in an individual's response to stimuli after it is repeated a number of times. A good illustration of this is when you first touch a turtle, its immediate response is to withdraw within its shell. However, after touching the turtle repeatedly, it soon realizes it is no longer in danger and does not try to hide. Even with new pets, it takes time for them to become familiar with you and your intentions toward them. Showing loving concern to them by taking care of their physical and emotional needs creates a pattern of familiarity with something pleasant and thus instead of trying to hide from you they are drawn to you.

Although I'm not trying to imply that the Sasquatch are like animals but this *is* the basic concept when applied to our forest friends and our attempt to bring them closer to us. The old way of trying to find them by what I would call the "Stalk and Gawk" method has yielded very little success over the past half century. In this case, a research team goes from area to area in pursuit of the "elusive" Bigfoot using the latest of detection gadgetry such as infrared/thermal cameras, parabolic dishes, audio recorders, etc. When they get no results from all this, they will quickly conclude that Bigfoot is nowhere to be found and move on to a different area. Of course, most of them will reason that they are dealing with some kind of large ape-like animal and the thought that these creatures are even remotely capable of outsmarting them, is sheer lunacy. Aren't humans the smartest beings on the planet? ...and how dare you think otherwise! People who think this way will be in for a big shock and wake up call to what the reality truly is!

Unfortunately, it is still the method that most old school or traditional researchers try to use as well as the producers of numerous TV shows where they "hunt" for Bigfoot, even if the weapon of choice is a camera or some other electronic device. One would think that some of these folks would "get it" and understand this will get them nowhere. There is a popular saying that would apply here, "If nothing changes then how does one expect to get different results".

The thing one has to do is to first change their mindset about *who* and *what* the Sasquatch are. If one sees them only as animals who roam the forest and can be captured as such then he will probably never see one unless one of them feels we need to be "persuaded" to think otherwise! Based on the successful interactions different people are having with the Sasquatch... including me, we have to acknowledge that they are sentient beings who are just as intelligent as we are, if not more. We also have to put aside any prejudices about the way they live... just because they are hairy creatures who do not wear clothes and who chose to live "in the wild" without the need for manufactured material things including homes, cars and all the other so-called "necessities" we have as humans does not mean they are unintelligent.

Remember it wasn't long ago in the past few centuries that the so-called "civilized" Europeans who came to America viewed the Native American Indians as savages. Why? Because their culture and way of life was little understood by the white men. The Indians were thus labeled "uncivilized" and treated like animals as a result! The same could be said about how the native Africans were captured like animals back in the 1800s and shipped to the new promised land to serve as slaves. Even in our current times, there are numerous examples of one group of humans elevating themselves over others so everyone should be aware of what I'm talking about unless they are living under a rock!

It's sad to say but it seems the human race, in general, has difficulty in accepting others who look differently or whose culture, way of life or belief system is foreign from their own. This is no different than when we contemplate our view of the Sasquatch people (and yes I do mean people). Put yourselves in their "shoes" and reason out *why* they would hide from the human race in the first place. They are very different from us in how they look and in the way they live so it doesn't take a genius to figure out that they hide out of necessity... in order to survive! If they were to let down their guard as a people and started coming out openly in the public streets they would be hunted down and killed to extinction!

Thankfully, they have long been dealing at a distance with humans and have had plenty of time to observe us... to understand how we react to various things and our general temperment. They also have studied our weaknesses such as our limited eyesight and lack of strength (when compared to themselves). Whether they were created with special natural abilities or evolved that way to adapt to the environment, they do have a number of advantages over humans that allow them to stay hidden from us.

One of these abilities that many mainstream researchers refuse to believe is their ability to "cloak" whether by blending in with background colors or simply appearing translucent by somehow bending light around them. Another would be their incredible eyesight that allows them to see at night (without an external light source) just as well as in the daytime. They can also emit certain smells to ward off unwanted guests and from what I have personally observed, the ability to hide themselves in a covering of dark or smoke-light haze which is possibly one of the reasons for the many blurry pics people have taken. As if that weren't enough, there's the ability to "zap" someone using some form of electromagnetic wave or infrasound which we do not fully understand. This is probably the one ability that causes electronic equipment to go "dead" if we get too close for comfort.

Aside from their special abilities, they do have a cultural and clan infrastructure that they have to strictly adhere to in order to maintain their ability to live "in our backyards" without being detected. The young ones are taught from birth the ways of the clan and the reason why they must follow the rules! I'm sure you can guess what the number one rule would probably be: Keep your distance and avoid contact with humans at all costs! To me that would be a no-brainer with the second being: "If a human spots you, stay perfectly still until he/she passes by."

This makes me think of a funny experience of a friend of mine, Bob, whose first encounter with a young Sasquatch happened while camping. The little Squatch had come in a little bit too close while he was lying down. When Bob suddenly turned his head, he was face to face with the little one who "froze" in place - facial expression and all! He told me his first thought was, "Why is there a monkey smiling at me?" I still chuckle when I think about Bob imitating its facial expression! The Sasquatch took off once Bob turned his attention away from it momentarily.

Having an understanding of what I have presented thus far will go a long way in having the right attitude to making contact with the Sasquatch. One's motive for making contact also has to be for the right reason:

> "Am I looking for Sasquatch so I can get definitive proof that they exist and prove it to the world even making a name for myself in the process?"

This reason seems to be the main objective of many of the traditional/old school researchers. Are they even interested in getting to *know* the Sasquatch other than simply proving they exist? Or is it the *ego* that is getting in the way of what the true reason for contact should be:

> "Am I looking for Sasquatch so I can establish a long term relationship with them and learn as much as I can about their way of life and culture? Can the Sasquatch help us to better understand why we, as humans, are experiencing so much trouble in our adapting to life on planet Earth? Can we learn to adapt some of the Sasquatch's ways of life into our own so that we can have a less harmful impact on the planet?"

Granted the above rationale would be pretty deep and almost spiritual in nature but it is one that the Sasquatch would greatly appreciate as our geniune motive for contact. It is because of this that the Sasquatch are *allowing* themselves to be contacted and sometimes they are also taking the initiative to contact us – all due to our having the proper attitude!

Once we leave the human ego behind and develop the proper motive and attitude, we can proceed to learn more about the specifics of the habituation method:

Find a wooded area and look for evidence that they are actually there

The Sasquatch people have been reported all over North America, so a good place to start would be a forested area like a greenbelt that connects back to a primary forest or mountain range. I can only speak about what we found in the Pacific Northwest. In most cases where they have been seen, they seem to be either in the mountain ranges (including the foothills) or narrow greenbelts that ultimately lead back to the mountains. This, of course, is just a general statement to increase the odds of a sighting as, there are exceptions. I do know of a small quarter mile radius park right in the middle of the city where there is evidence of Bigfoot activity! Another important factor would be that the area has to have a natural source of water available like a lake, stream or river. All the different areas where we have encounters had that one thing in common; especially if fish were present, especially Salmon - a favorite food item for the Sasquatch.

Once one finds an adequate location, he will need to look for evidence like stick structures arranged in geometric patterns that could have only been made by intelligent creatures like humans or the Sasqautch. It's very seldom that humans will take the time and energy to go out in the woods to do this so it's important to check around the area and look for different structures. Large ones will pretty much rule out the possibility of direct human construction, as we are no match for the strength of the Sasquatch.

The most frequent sign seen is what we call the "X" structure and is made by pushing two trees over at a 45 degree angle and in opposite directions. Sometimes they will take large detached logs and jam them in the middle of other trees to form an "X" (see below).

Another popular structure are when numerous trees or logs are arranged to form a teepee. Here's an example of a giant one we found in the Blue Mountains of Washington:

Although a lot of these structures are erected vertically, they can also be constructed horizontally as was the case of this "A" structure we discovered in one of our most active areas:

Also if lucky enough, one may even find a structure built over the creek to form a bridge as we found near one of our gifting areas:

Visit the area on a regular basis

Once you have confirmation of the Sasquatch being in your area, you need to take the time to go there on a regular basis. Remember the definition of "habituation" where there is a decrease in an individual's response to stimuli after it has been repeated? Well, by your frequent visits to the area, the stimuli, the visitor, will decrease the Sasquatch's automatic response to his presence. In other words, the visitor will become less of a stranger to them over time to the point where his being there is almost expected by them!

What does one do during these visits? He will need to take the time and be very observant of his surroundings. He needs to listen and see everything around him. He should take note of anything that could pose a danger to him being there such as certain wild animals that may frequent the area such as large bears, cougars and wolves to name a few. In the particular area I frequent here in the Cascades, there is little danger from these although I still keep a watchful eye while I'm out there just in case.

Should a weapon be carried for protection? That's a personal decision and one where I cannot suggest one way or another. In *my* case, I do not carry any because the clan here is very sensitive about armed individuals walking through their home. It even unnerves me to come across someone carrying a weapon. One day, someone I know had a bad experience with a guy who was carrying a rifle out in the forest. The guy got spooked as he walked by because my friend was off the trail, partially hidden behind a tree and was seen in the corner of his eye. It startled him so much that he momentarily pointed the rifle at my friend until he realized he had it pointed at an unarmed individual. Long story short, I have nothing to fear from the Sasqautch – it's the humans who I worry about!

Once familiar with his surroundings then he should find a spot to hang out. Bring a lunch and enjoy the nature around you!

Demonstrate your respect for the area

While in the chosen spot, one needs to demonstrate respect for the land by not littering. If something is brought in to eat or drink, make sure to pack the garbage back out. Remember... the Sasquatch have had plenty of time to observe the worst of humans; the garbage left behind in camp grounds with beer cans, bottles and other kinds of trash. They will have nothing to do with people like that so we need to stand out as being much different from the general herd. It's also a big plus to them if, in addition to not littering, we take the time to clean up other people's messes!

One will need to behave oneself while there. This means not being rowdy, noisy or disorderly by yelling, screaming or banging on trees with a stick to get their attention. Believe me, the Sasquatch will know you are in their home the minute you set foot on the trail. You should also avoid disturbing the area by breaking or cutting trees or the branches. Also avoid changing or removing structures they have built... they are there for a reason even if we don't understand that reason.

Start a gifting spot

Setting up a gifting spot is a great way to encourage interaction with the Sasquatch. Finding a large stump left over from logging operations would be a nice spot to place gifts. So what kind of gifts should be left?

The first thing that comes to mind are food items such as peanut butter and honey. We have been successful with food similar to this but there are some precautions that should be taken before leaving food as a gift. Remember there are other creatures in the forest who would love to get to the food gift before the Sasquatch have a chance to check it out. Therefore, we need to make it hard for racoon, deer, elk and other other animals to get to the food by first off leaving food in sealed jars with metal lids and preferably, glass containers. Depending on how aggressive the local animals are, you may need to place your food items in a bucket and suspend it up high in the air with a rope thrown over a tree limb. This is no different than the precautions one should normally take if he were camping and needed to safely store food in the open. Also, we do not want to run the danger of being accused of illegally baiting animals so be sure to check your local laws to make certain you're in compliance.

Likewise, never refer to food gifts as "bait". Not only is this disrespectful to the Sasquatch but humans may get the wrong idea about what is happening out there in the first place. Do not worry about making it too difficult for the Sasquatch to get to your food gifts. They are intelligent people like humans and will have no trouble figuring out how to open jars or getting things out of a bucket in the air!

As for non-food gifts, try almost any small item that may attract their attention. We have personally tried using polished gem stones, small toys and stuffed animals like a baby bear.

It's important to note *how* these gifts are arranged (including any food items) on the stump. Taking a picture after everything is set up will help later on when we return to see what moved or is missing. It will reveal what interests them individually and what kind of food they like to eat.

If the Sasquatch partake of the gifts, they will no doubt express their appreciation by leaving gifts of their own on the gifting spot. This might be anything in nature like flowers, leaves, bones, stones, shells or special shaped wood roots. Since Barb does gifting much more than I do and the local clan have come to know where she lives, they have left numerous gifts at her place. One outstanding example was a rock left in her yard that looked like it had gold paint sprayed on the side of it. When she had it checked out by someone, it turned out to be real gold flakes coating the rock! Sometimes the gift can be as small as a leaf. After we had returned from our trip to the Blue Mountains, Barb went up the trail to our habituation area to talk with the local Sasquatch about our experiences. When she returned she immediately noticed this one little leaf stuck in the door handle! It is very unlikely that this happened naturally.

Screenshot from Barb's Video showing where the leaf appeared

Play a musical instrument

To really get some great interaction with the Sasquatch, bring a musical instrument to the area. In my case, I started by bringing my mountain dulcimer and played it for fifteen minutes. The stick snaps and knocks would increase, not only while I played, but especially when I stopped. It's like they were telling me, "Don't stop now! Keep playing!" I have also used a plastic recorder flute and as of late, started playing a hand-made wooden Indian flute.

By regularly going out and playing music, the Sasquatch will get in the habit of hearing it and will miss it if it is not kept on schedule. Don't worry if it is played badly... they appreciate it if it comes from the heart. Once again, I personally feel that we have to be respectful when it comes to how loud we play or if we intend to use a boom box to play pre-recorded music. Playing harsh, heavy metal music might attract some of the young ones as it would any human teenager, but it probably won't sit well with the clan as a whole. So it's best to keep it simple and pleasant.

Pay attention to changing stick glyphs and structures

After being at the area repeatedly, one should be able to take note of any changes to stick structures and also sticks along the trail that change or are added. In my and Barb's experience going up the trail to our area, we have found patterns in the way small sticks are layed on the path to form language characters called glyphs. Some of these are quite simply such as "A","H" or "X" but we have also noted more complex ones that look like Chinese characters. The key characteristic is that they are intelligently arranged and not just a bunch of sticks randomly falling on the ground.

When observing these small glyphs, take note of *how* the sticks are layered. Are different sticks layered in such a way where one part of the layer goes back under another? This does not happen naturally and is a sure bet that this was created by an intelligent creature.

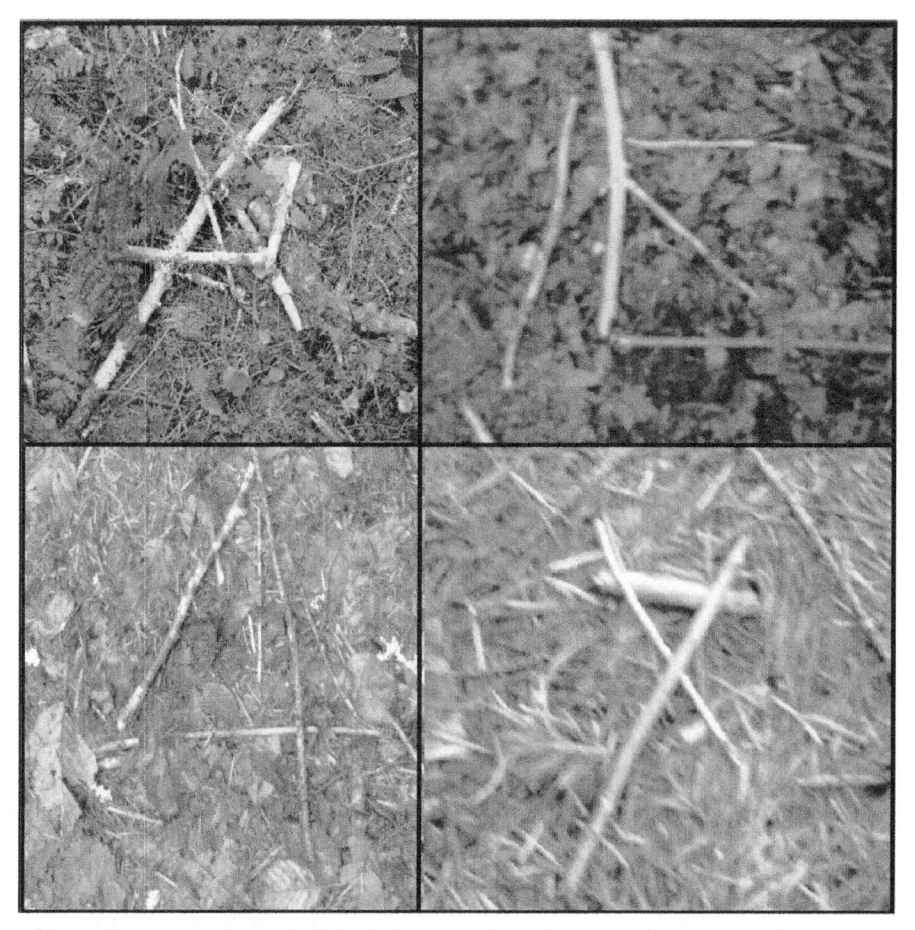

These are some of the stick glyphs we found along the trail in our habituation area. Notice the various geometric patterns.

Learn to listen carefully

The other important thing one needs to learn to do is to listen carefully each time he is in the area. Have the sounds of the birds changed? Are certain animals like chipmunks making a racket for no apparent reason? You can learn a lot from this as the birds and animals will act as warning signals of something approaching the area whether animal, human or Sasquatch. Tune in to the smallest of sounds and you may even hear the Sasquatch whispering or humming to each other. The chapter in this book titled "Stop, Be Quiet, Listen!" will give more in depth information on this subject.

Although I have a good eye for spotting them in photos, my hearing is a different matter. One of my friends, Penotia Sesquai, an audio analyst from Alaska, has a YouTube channel and will examine other researchers videos to reveal what she is able to hear. What she can hear is amazing whereas I can only pick out bits and pieces of what the Sasquatch are saying and yes, they know English (at least here in North America)! Check out the videos on her channel – look for the web link on the Online Resources page of this book.

Do not leave recording devices other than what you personally carry on you regularly

This is something we found out by experience with leaving audio recording devices at our gifting areas. Often, we would turn on the device to record overnight and come back the next day just to find it shut itself off mysteriously within minutes! On one occasion, we decided to hide one of those spy USB recorders under a bunch of leaves on the tree stump used as a gifting spot. Later, when the Sasquatch showed up, they immediately *knew* it was there, dug it out and it ended up fifty feet away from the gifting spot! We know this because we eventually found the device and could hear them grabbing it and then all was quiet. According to Barb, they never touched the gifts after that episode but can't say for sure if it was due to the spy device being there. My personal feeling is they don't like being deceived or treated this way. How would *we* feel if someone placed a hidden spy recording device in our home somewhere? Wouldn't we feel like our privacy had been violated? This is no different for them when we go into our habituation area because in reality it is *their* home.

However, it's different when we carry our audio recorder or camera *on* us and out in the open at all times. The Sasqautch can see we are not trying to hide something from them and after a while, they will get use to what we are carrying. In my case when I started, I had my cell phone camera with me all the time and I used it to take landscape photos of our area. Later, when I obtained a professional camera, I took it with me as well. Barb also has her Ipod attached to the top of her walking stick and always has it on record mode while out there. As you can tell from the various photos in this book that I have taken, the Sasquatch have no problem showing up somewhere within a landscape photo. Especially if I have done what I have already suggested in this chapter, they will have a level of trust with me and will not try to stay completely hidden.

What I just mentioned works in the daytime. However, at night, I would strongly advise not using an infrared camera. Even though you can't see it, most infrared cameras require beams of infrared light to shine out in order for this type of light to be reflected back into the camera (that's why you will see a number of LED lights surrounding the lens of these kind of cameras). The Sasquatch can see in the infrared part of the light spectrum and it will look like a flashlight shining in their eyes – something we never want to do!

A thermal camera operates differently (and can be very expensive) by detecting what's below the light specturm... heat. Since all things emit varying amounts of heat even in the night, the thermal camera does not need to emit anything – its job is to simply detect heat coming from wherever it's being aimed. We do have a thermal-camera that we use use occasionally but even then, we have had it mysteriously shut off. Barb did record what looked like small Sasquatch coming out of the forest into the street for a few minutes but other than elk and the image of Gabby coming up the road toward me and giving me a scare (I thought she was a bear!), it has largely been unused. My conclusion on the thermal is if the Sasquatch prefer to not be seen at night, they will let us know by temporarily disabling it.

Remember we are in their home and as such, need to act like a guest

This just emphasizes what I had been saying. Think in terms of what we would expect from others coming to our own home as visitors. We would always want our guests to respect our space by not walking into our house at all hours of the day or night and having a rowdy party with a bunch of folks we don't know and throwing garbage all over. It's even worse when the guests decide to carelessly damage our house or possessions to the point where our home is unlivable!

Just compare this with the general human population's attitude toward their environment. While a lot of people would feel angry if what I just described happened in their home, many of these same people would not hesitate to throw garbage out of their car windows while driving or dump garbage while camping in a beautiful place. This has aways struck a nerve with me since I was young. If one recalls in my childhood years, I would pick up the trash left by others in the woods behind my home because I cared about nature. Not to pick on my deceased dad but I even got upset when we went to the drive-in theaters of the day and he would throw the popcorn boxes and cups out the window right before we left. His reasoning was, "They have employees that will pick this stuff up" but of course I still disagreed. I knew that *anywhere* on the Earth is by extension our home and we needed to respect it.

Being that the Sasquatch generally live in the forests and see this happen to *their* homes on a regular basis, it's no wonder why they feel contempt whenever they see a human. That's why we have to do our best to show them that WE are different from what they normally perceive of humans. Not only do we NOT leave garbage, we should make the effort to clean up the area of what others leave behind. Also, when it comes to the trees and plants, we want to show our respect by not breaking or cutting live branches just to clear a path. Something we noticed is, if the Sasquatch have bad experiences (or are uneasy) with people coming down a certain trail near to where they hang out, they will purposely push trees over or place sticks or logs over the path to discourage future intruders. Clearing a path by cutting through what they placed there as a deterrent to humans will not earn you favor them. Just leave everything in place and do not disturb it. Showing them this kind of respect will certainly go a long way toward you being a friend of the Sasquatch. They will certainly notice it and will feel a great deal of appreciation for your concern of their home!

Be the kind of person that would attract the Sasquatch to you

I'm a big believer in karma and the law of attraction. If one is a positive individual then others who are also positive will be attracted to you. On the flip side, if we dwell on negative thoughts or watch negative TV programs or movies that focus on hate or fear then we will attract those same negative energies back at us. This is very similar to what we called the Golden Rule and is also a Bible principle: "Do unto others as you would have them do unto you ".

This is something I have worked at all my life. I always assume the Sasquatch are observing me even when I'm not in the forest. In the course of my day, I try to be the best person I can by being caring towards others, honest and ethical in all things. I don't participate in the drama that consumes a lot of humans and I *do* keep my distance from those that do. The Sasquatch can "see" our attitude emanating from us as either negative or positive energy. So in essence, we have to show ourselves worthy to have contact and a relationship with them. Most traditional/old school researchers who look for Bigfoot will never see this connection between *who* we are as a person being the determining factor as to whether or not we will have any contact with them at all!

Be patient! It takes a while to gain their trust of you

There is a saying, "Rome was not built in a day" and neither should you expect the Sasquatch to warm up to you overnight. This is a process that takes time. This is the same thing when we meet a stranger for the first time who is being friendly toward us. We would be very uneasy with that person if they came in too close to our space. Not knowing the person's intentions, we would be somewhat guarded and limit our contact in the beginning.

However, as time passes, through the process of habituation, we become more familiar with new people we meet: how they act at different times with other people (basically observing their temperment). We will also observe how others are treated by them and learn a bit about their background: where they are from, what kind of work they do, what kind of hobbies they enjoy, etc. At the same time, they are also sizing *us* up. Based on ongoing interactions, we will either become closer to them as good friends or it could go the other way where we keep a distance from each other because of mistrust or a person's unethical or backstabbing nature.

So give the Sasquatch a chance to get to know us during the

time they are observing. Be the kind of person that would attract them to you and let nature take its course. Don't be discouraged if days, weeks or months go by with little or no interaction from them. Try to focus on being a friend of nature itself and feel the love that comes from everything around us in the forest. Talk to the Sasquatch, even if you don't think they are there and express a appreciation for being in their home. Express gratitude for the beauty all around. Those are some of the lessons the Sasquatch *want* us to learn… it's all part of the process of getting to the point where they can be closer to you!

    I can honestly say that I have been personally blessed with numerous interactions with the Sasquatch. I'm always happy when I return from my habituation area after spending time talking with them about my feelings, taking some landscape photos and later discovering they were nearby listening to my every word! As time goes on I feel assured that our interactions will continue with greater meaning!

Dr Matthew Johnson and his partner Cynthia Kreitzberg

Dr. Matthew Johnson, a Licensed Clinical Psychologist in Washington and Oregon, has been very successful in using the habituation method to get to know a local clan of Sasquatch down at a remote location in southern Oregon. The spot is known as SOHA (Southern Oregon Habituation Area) and for the past ten years he faithfully would go there from Washington state to stay for a weekend each and every month. At first, staying in a tent and using gifting bowls of different food items placed on the perimeter of camp, the Sasqautch (especially the younger ones) eventually got used to Matt being there and started taking some of the food from the bowls. Eventually the adults also started coming in closer to keep a watchful eye on their young as the interaction increased between them and Matt. He has documented many sounds made by that clan including very human like speech. Matt, as well as his partner Cynthia, have also had multiple sightings of them around their camp

Over the past several years, Matt has invited others to join him on these trips to SOHA and most can attest to the experiences of not only hearing the Sasquatch but also witnessing their other unusual abilities such as cloaking and even what can best be described as a light portal that first opened up one night during 2014 while two researchers (Adam Davies and John Carlson) were with Matt at SOHA.

I would also like to mention that Dr. Johnson has been very instrumental in my becoming a researcher and his experience and insight was the needed push to get my research out to the public via YouTube, Facebook and now by means of this book. Both Matt and Cynthia are good friends of mine and I will continue to follow their adventures in Oregon with great interest!

More information about Dr Johnson's experiences with SOHA and how he can be contacted can be found in the Online Resources page near the end of this book.

In conclusion, what kind of results can one expect when using the habituation method of getting to know the Sasquatch? The photos presented in this book thus far and the following additional photos I have taken over the past year will speak volumes in answering that question.

"A picture is worth a thousand words!"

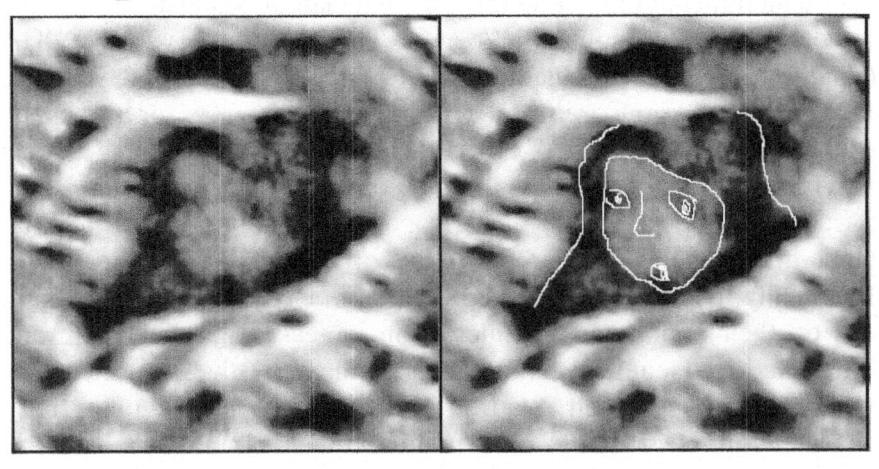

Here is a very young Sasquatch that was in the Field of Dreams habituation area this past July 2016. I have a comparison photo of this little guy in my chapter "Photographic Evidence".

A juvenile Sasquatch popping its head out from the side of the tree (center of picture). Photo taken in the foothills of the Cascades on January 9th, 2016

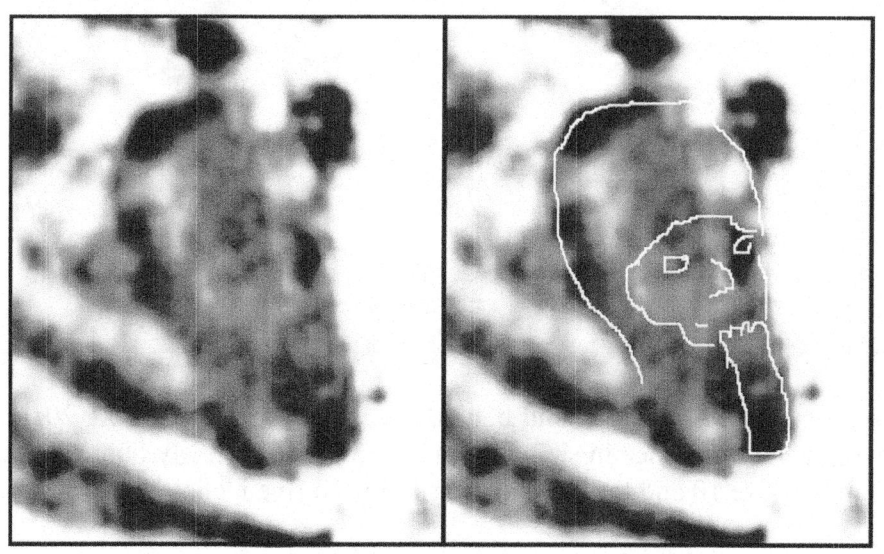

On April 8th, 2015 I took a photo of a ridge where the Sasquatches frequent after playing my flute/recorder for a bit. I discovered this baby with very distinct facial features up in the tree.

A single footprint was found behind the local tavern on Jan 10, 2016. Notice the difference in size with my shoe heel to toe. My boot size is a men's size 10 and the print is clearly four to five inches longer.

This is a younger Sasquatch (behind the main branch) that was sitting across the creek from our gifting spot. I call it the "Relaxed Sasquatch" as its left arm is behind its head while its right hand is holding something with a white top. This is another example of enlarging a small portion of the main photo in order to see the details more clearly.

On December 1st, 2015, a fellow researcher accompanied me to the area where I had witnessed the strange UFO manipulation by John back in September. It was the first snow of the winter and we wanted to check out the structures further down the old logging road. What amazed us was the number of trees and branches that seemed to be piled along the way to purposely discourage anyone from going in that direction. Coming upon the first piled structure, I took my first photo before continuing on to where some large X and A structures were found. When I had the chance to enlarge the photo, I found what appears to be a smaller Sasquatch sitting in the back, off to the right side of the wood pile.

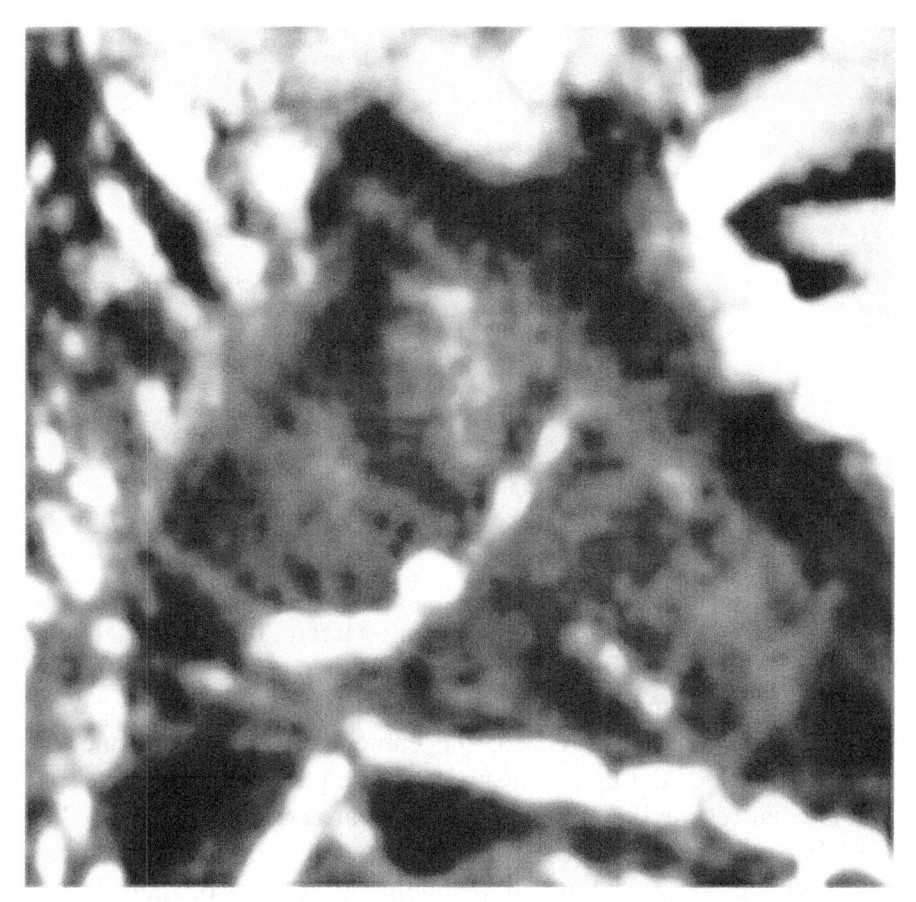

Here is a close up of the Sasquatch from the previous page. What is so amazing here is the facial detail of near human features. Though the body is covered with hair, the face is rather clear with its right eye, nose and mouth being completely visible. I don't see any expression of fear though I'm being stared at rather intensely as I'm taking the photo. Although it may be difficult to see, a baby is being held on its right side as one can see its small face looking sideways.

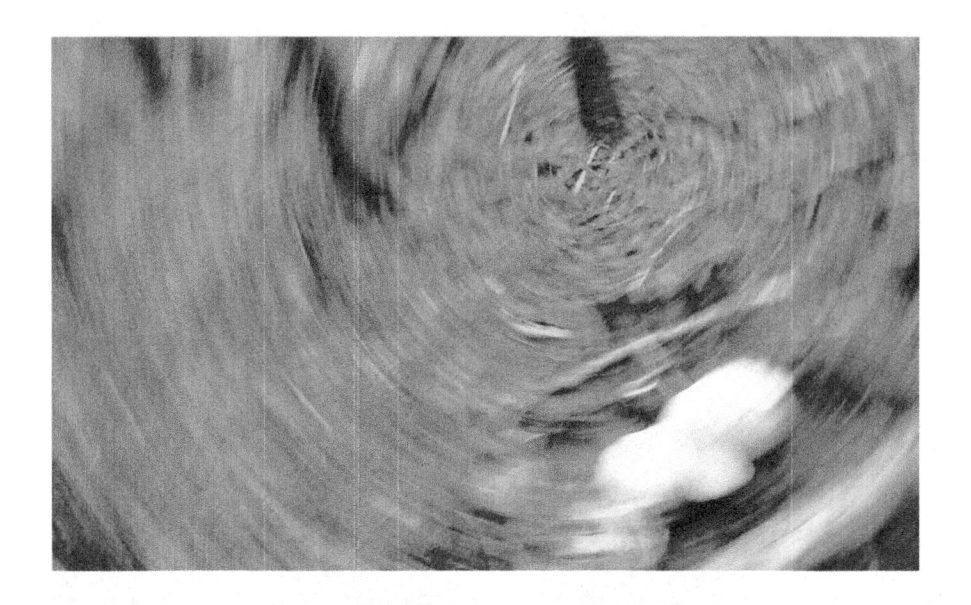

I had returned to our gifting spot in mid-March 2016 and discovered that a sealed honey bottle had been taken after being left there a few days earlier. I wanted to take a photo of the spot for comparison purposes but, for some reason, the camera on my cell phone was acting strangely. As I held it steady to take a photo of the gifting spot, this is what resulted! Astonishingly, there are certain objects, such as the white stuffed toy, appearing stationary while the rest of the photo is caught up in some kind of magnetic vortex. Also, there are no special effects like this available on my phone that could possibly cause this.

As a final note I have high quality color versions of these photos in my Photo Gallery of my web blog site **http://planetsasquatch.com** for one's closer inspection.

Chapter 21
Photographic Evidence

Based on all the questions I've received from viewers of my Planet Sasquatch YouTube channel, the one that stands out from the rest is, "Why do your pictures of the Sasquatches seem a little blurry most of the time?" I felt that this question deserved a thorough answer and so I devoted a chapter in this book describing the issues of taking pictures with respects to our forest friends and the some methods I use to produce the pictures I currently have.

The first thing one has to realize is that the Sasquatch will never get close enough to you to have a selfie taken with them. Most times, when they're nearby, we will usually never realize they are there. So it's pointless to think we will know precisely *where* we will aim the camera when we do not know precisely *where* to look. I know this may sound like a no brainer but it is surprising how many people that think the Sasquatch was standing right in front of me or was clearly seen nearby. Some of them also think that I would get better results if I used a more powerful telescoping lens. That would be great if, once again, I knew exactly *where* the Sasquatch was in the first place.

If one is successful in his habituation efforts, he can rest assured the Sasquatch are watching from somewhere in the landscape, especially if he hears stick knocks and sounds mimicked by them. Hence, taking pictures of the surrounding area as landscape photos should not be intimidating to them especially if this is what is done on a regular basis. Chances are, he will not be pointing the camera straight at them but they will inherently be included in the landscape with a wide angle shot and usually no telescoping needed.

Somewhere in that photo, whether partially behind a tree or bush, there's a good chance to spot them. However, *where* they appear may be in a very small portion of the full picture and will be much too small for one to make out any detail without first enlarging the part where one suspects seeing one.

This is where the power of the computer comes into play.

Most cameras nowadays are digital and easily loaded onto a hard drive. There are many software programs out there that allow the viewing and enlarging of the image on your screen. A basic one on Windows based computers is its image viewer and this was the one I started out using. This allowed me to explore the entire photo piece by piece looking for what could potentially be a Bigfoot. Sometimes by enlarging a photo this way, you can also confirm what you thought was a Sasquatch but turned out to be a tree stump instead.

Let's say that the photo was enlarged and we spotted what looked like a Sasquatch without question! Well, one would be surprised at the number of times that I had, what I thought, was the real deal but in going back out to the spot where the photo was taken, it turned out to be part of a tree or stump. Hence, the importance of printing out a copy of the original photo, circling the spot in question, and then going back out to the spot with the printed picture in order to take another identical photo. It is necessary to find the exact spot where the photo was taken in the first place. By carefully examining the branches of the trees in the printed photo, we will be able to line up the exact angle of how the branches cross each other to find the spot we're looking for. We may have to adjust any telescoping lens in order to match what is seen in the camera with what's printed in the original.

Once the comparison photo is loaded in the computer, it's very easy to pull up both photos at the same time in the image viewer and adjust the enlargement until the background matches. If the Sasquatch is still there, especially if the second photo was taken the next day, then chances are this is some stationary object and will require a return trip to examine the spot directly for a final determination. It would be fortunate if the post picture showed only vegetation and was missing the object of attention in the original!

In Chapter 9 "Direct Sasquatch Interaction", I had the two cell phone photos of the large Sasquatch taken a few seconds apart. Although I could see it had moved between the photos, I felt it would be more credible as evidence if a third picture was taken of the same spot but this time, on a different day. After lining up all three photos to the specific spot between the two trees, I could make a reasonable claim that there was a large Sasquatch standing there and not some large dark inanimate object. Had the third picture also yielded a large dark spot then, chances are, this would have been some kind of shadow anomaly.

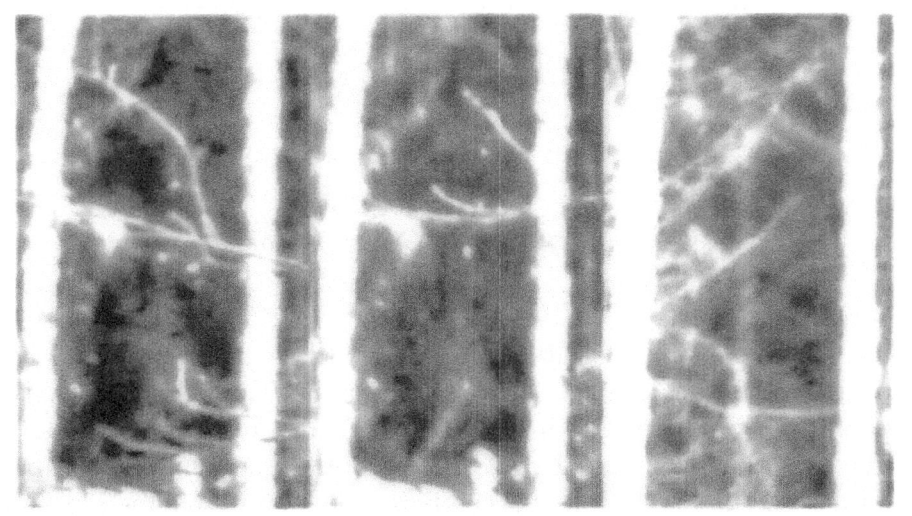

I have cut and pasted the targeted area from all three photos and arranged them side-by-side as shown above. As one can see, all three photos have the same two trees showing up and at the same height and distance. Notice how we can see the small tree in the middle on the third photo. However, in the first two photos, it is completely blocked, as well as the other surrounding branches, by the dark massive body. Of course, being able to see the shape of the body, especially the head and facial details, makes the evidence that much more credible.

Another example is from one of my first photos taken with my Nikon camera, what I call the "Creekside Squatch" on my YouTube video. The photo at the top of the next page shows a domed head with two visible eyes. One of the things the Sasquatch do, in trying to remain stealth, is to position themselves in front of tree trunks. This makes it difficult for onlookers to distinguish between the tree and a possible Sasquatch - just enough to make you question it and perhaps overlook it.

If you look closely, there is, in fact, a tree trunk directly behind the Sasquatch. The facial detail warrants further study of this photo and to determine, whether or not, it is part of the tree after all. The first clue was the color and shading of the tree and that of the Sasquatch. In looking at the color version of this photo, I could identify the subtle differences in grayish/brown color between the tree bark and the Sasquatch, as well as the texture patterns since bark doesn't usually look like hair. The next step was to bring the printed copy of the photo back out into the field, in order to take the needed comparison photo.

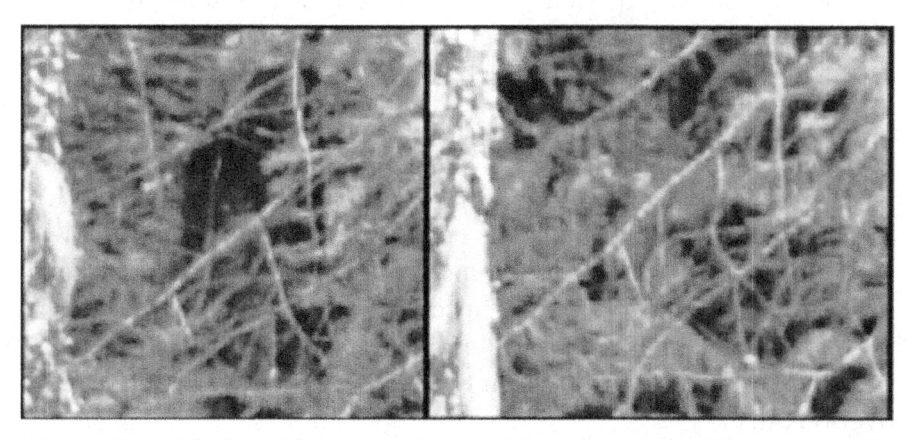

Above is a side-by-side comparison with a photo taken later from the same spot. In doing these comparisons, I try to line up the tree and branches as closely as possible. The photo on the left shows the young Sasquatch, whereas the one on the right just shows the background, confirming this was not a tree trunk with a face, unless it somehow unrooted itself and walked away.

The photo below is an enlargement of the original to show more detail of the facial area. The two eyes and pudgy nose can easily be seen:

Below is another example of a comparison photo taken a bit later from the original back in July 2016. A close matchup to the ferns and branches help to identity the exact spot where the young Sasquatch was.

Another important consideration is the camera's ability to produce high resolution photos. Resolution has to do with the amount of detail the camera is able to capture and store, especially if the photo is later enlarged.

Notice that the above photos seem identical with the small, white object on the pole in the middle. The top photo was taken with my cell phone's 8 Megapixel camera while the bottom photo was taken with the Nikon D3300 24 Megapixel camera. The differences become apparent when the photos are enlarged and compared using the image viewer.

| 8 Megapixels | 24 Megapixels |

After enlarging both photos in order to zoom in on the object, we can tell by the photo on the right that this is actually a stuffed Fairy Elk toy. The detail is still evident, i.e. the face and eye, as well as its furry body. However, the cell phone photo on the left has already started to pixelate, small squares appearing due to the lack of digital detail. The further you enlarge the photo, the worse the resolution becomes. Therefore, resolution simply refers to the number of pixels that a digital photo contains and this, in turn, directly affects the clarity of the image.

8 Megapixel	24 Megapixel

Further enlarging the photos make it very evident concerning the differences in resolution. At this enlargement, the 24 Megapixel photo on the right can still be discerned easily and is just barely starting to pixelate.

So investing in a good high resolution camera will go a long way in aiding your research. My recommendation is to obtain, at least, a 24 Megapixel camera similar to the Nikon, which is still reasonably priced. If you have a few thousand dollars to spare, you can go up even higher to 36 Megapixel. As time goes on and technology continues to advance, expect camera prices to drop while offering better features and higher resolution such as 50 Megapixels and up.

One final thought on cameras: Become very familiar with your particular camera and especially with regards to the focusing features. Since we're discussing taking landscape photos, it is important to adjust the camera's focus for the furthest object in the background. Nothing's worse than taking a landscape photo with nearby branches being crisp and clear and the background being blurry.

Chapter 22
Communicating with the Sasquatch

There are several ways in which the Sasquatch peoples communicate with humans. Those of us who are privileged to participate in this communication can best describe how it takes place. There are various levels of this interaction starting with a feeling and progressing over time to formed mental thoughts/feelings and then translated to words that most humans will understand. Personally I have developed some psychic/empathic abilities over the past few years and am at this particular level with my interactions with them (and growing stronger with each passing day!). Some, of course, have already advanced to communication with them involving specific messages being mentally received complete with names and places being projected. I would definitely call this the "mindspeak" level and I know of a couple of people who have regular conversations with them.

All of this can be thought of as telepathy as opposed to channeling. Channeling is a method of communication where an individual takes a back seat consciously while an entity takes control by doing the speaking directly using the host's own vocal cords. The channeler simply becomes a conduit for the entity, much the way a radio receives messages from electromagnetic waves and converts it to sound waves. The channeler may or may not be aware of what the entity is saying once he or she regains control of their mind and body.

Telepathy, on the other hand, is simply another form of communication versus using the vocal chords to create sound waves. Although it is not fully understood, telepathy is a mind-to-mind transfer of thoughts usually in both directions. This may be one of the methods of communication between animals and birds aside from the sounds they make or possibly used in tandem. This is also true with the Sasquatch although they do have a spoken language.

Humans, on the other hand, seem to be severely lacking when it comes to this ability or so it seems. As will be discovered in a later chapter, "Human's Artificial World", there are reasons why we have been disconnected from being able to use telepathy with each other and other life forms. I can say that this state of disconnection was by design in order to control you!

All is not lost and people are waking up and taking steps to regain these abilities such as telepathy and even telekinesis, the ability to mentally move objects without physically touching them, as was the case with John mentally manipulating the UFO. Later chapters in this book will give insightful information that may help put one on the path toward this goal of telepathic communication, as well as developing a trusting relationship with our Sasquatch brothers. As an empath, I *feel* the Sasquatch's desire to communicate with those that are ready. They need us to communicate the messages to other humans because of their difficulty in putting into our words the exact thoughts they are trying to get across to us. They know that if they can project their thoughts and feelings to a receptive human, the human can then draw on his or her experiences of living in the human domain to complete the expression. We can take their thoughts and frame them in sentences in such a way that the maximum impact can be received with one hundred percent comprehension. This is how it works with me and it is not channeling. It is real communication between me and them collectively. Their thoughts become my thoughts as I meditate on important issues. Once a connection is made with them, it does not matter where you are – they basically have your number and they can initiate the communication.

Therefore, the following chapters are really a combination of their thoughts to me as well as the thoughts of my higher consciousness working in unison with their messages (it was my spiritual guides that lead me to the mountains in the first place to seek out the Sasquatch peoples in order to connect with them during this critical time). I feel their frustration with the situation today and it leaves me with this burning desire in my soul to get these messages into impacting and meaningful words that conveys exactly what they want us to know.

Chapter 23
Stop, Be Quiet, Listen!

This was one of the first thoughts I received while in the forest. What is the significance of being quiet? The purpose is to clear our mind of the other clutter that typically fills most human's minds throughout the day. One of the reasons the human race is having tremendous problems is they take in voluminous amounts of information each day... almost a hundred fold compared to a century ago. This information is a collection of conflicting clutter that consumes peoples attention to the point where they can no longer act independently of what is being drummed into them from the outside. Most people claim they have "free will" but do they really? Are their desires and resulting actions not shaped by continuous attention to certain types of information?

Look at the world today as compared to a century ago. In those days humans actually spent more time focused on their work at hand and had only two real mediums of information exchange: direct oral communication and the printed page. Even so, most conversations took place at the dinner table and newspaper/book reading was done while relaxing, especially during long cold winter nights. Today this has all changed. Now, instead of two mediums, there are now many: television, radio, computers, and smart phones to name a few. The amount of time spent using these mediums has also grown to such a point where people now take in information on a continuous basis from dawn to dusk.

The dawn of the internet took this to a whole new level by allowing the same information (whether true or false) to be shared among millions of people. Whether or not people realize it, the constant intake of information will program you, shaping your desires and opinions, as well as taking sides on issues such as politics and religion. The negative outcome of all this is the dividing of people on trivial issues that serves as a distraction away from more important things like taking proper care of our planet. Hence, it makes it easy for a few to control the mass population simply by controlling the media of outgoing information. So the "will of the people" now becomes "the will of the controllers" and true free will is thrown out the window.

There is another negative aspect to all this mental overload coming in from the outside: it drowns out the voice that comes from within each person. I am not referring to something you hear with your physical ears or any other external input you may receive with your other four senses. The voice I'm referring to is the invisible communication medium that allows "free will" thoughts to enter into your mind independent of and not controlled by the external "noise" mentioned above. Imagine someone who constantly goes to a Heavy Rock concert and sits near the massive speakers used at these events. After the concert is over, try talking very softly to that person and see if they actually hear you. Your mind works the same way. If you blast information continuously through your physical senses, you will become very desensitized and not be able to mentally "hear" or focus on your own personal thoughts. As a matter of fact, it is like an addiction for some people to have continuous input like listening to the car radio or using a smart phone to the point where they will go crazy if they're not "plugged" into something to keep them occupied.

What happens if we do unplug from this massive stream of mental data? The noisy clutter that filled our minds will slowly dissipate and eventually allow that small "voice" of the real us to start communicating with us. In the beginning I may say, "I'm not thinking of anything!" but in time, free will thoughts will begin flowing and we will know it. This is what meditation is all about. It is the clearing of one's mind of all the concerns of the day and allowing a state of consciousness where we ponder the greater meaning of life. Our emotional state is important while meditating as it is impossible to do so while experiencing anger, sadness or any other low vibrational feelings. From personal experience, the best emotional state to be in, would best be described as being calm and living in the moment. It's a state of mind developed over time and once I'm there, a wealth of internal knowledge will be available!

So "Stop, Be Quiet, Listen!" was an important first lesson the

Sasquatch taught me to allow me to advance in my interactions with them. In the beginning I only applied this to my time in the forest… being quiet while I carefully listened to all the wonderful sounds made by birds, animals, insects and the occasional Sasquatch knocking on a tree. Being empathic, in time, I started to sense from them that this "quiet time" had a greater meaning and was really about getting me away from the damaging effects of the mental "noise". I got rid of cable TV two years ago and restricted my online intake of information to specific topics that were not part of the collective programming of the masses. I increased my time out in the forest while greatly reducing time in the city. I was able to develop a meditative mind simply by being in quiet places. Going back and forth to the nearby town, a twenty five minute drive, I found the car was an excellent place to meditate… especially when there is an awesome view of the mountains while driving. Calm, tranquil music was very helpful in the beginning although now, I like meditating without music.

The other distraction that needs to be avoided in order to meditate is being around people, especially if they can be heard talking nearby. We need to have quite a bit of "alone" time away from others for meditation to be effective. After a while, we will look forward to this time alone and, being around someone constantly, will start to bother us (not necessarily a bad thing – everyone needs some alone time). For example, as of today, I can only be around a large group of people for a short time due to the "overload factor" especially when I can hear all of them talking.

In conclusion, I now take a great deal of delight in quiet meditation where ever I am. Thoughts now flow to me, teaching me many things and providing answers to many of life's questions. If there is a key lesson to be had by reading this book, it would be to learn the one important lesson of this chapter, of establishing for ourselves the communication with our inner being!

Chapter 24
Human's Artificial World

One of the reasons that keep us suppressed is the fact that we, as humans, are the only creatures on this planet that require artificial structures and manufactured material things in order to survive. Think about it – every other creature that lives on this planet is well suited and well adapted to the Earth's environment *except* for mankind. We seem to require these foreign jungles of concrete, brick and mortar which separate us from the natural world around us. Most people, especially those in the cities, live their entire lives with limited access to nature and are, in effect, cut off from the benefits of being fully connected to our planet such as the Sasquatch and just about everything else. Someone will argue, "But we can't just simply live out in the woods, forging off the land as the animals do!" Well that's true... now... at this point in time. However, I would then counter with, "And how do you think we got to this point of dependency in the first place!"

Our idea of human progress has been twisted from the very beginning of our being created as a hybrid race (yes, you read correctly). Instead of trying to learn to live on what the earth provided, using only what was personally needed for self, family or clan and giving back to the earth when it was no longer needed, we, as a race, opted to build our own artificial environment of massive cities completely cut off from the natural. What has been the result of this so-called "progress" to our planet? We have damaged all the ecosystems that are critical for the continuity of life. We have poisoned our air and waters with chemicals and, as of late, nuclear radiation (i.e. Fukushima, Chernobyl and others). We think, as humans, that we are the all-knowing beings who can make the best decisions when it comes to these ecosystems. Geoengineering has affected the natural weather patterns. Massive flooding is happening in one area of the earth and severe droughts on the other. The fish and mammals in the oceans are dying, the birds are dropping dead from the skies and everyone, humans, animals and our forest friends alike, are suffering from having their homes burned out by the massive fires that have raged across the planet. If this isn't bad enough, there is the destructive behavior of the ruling elites that seem intent on finishing the job of turning our planet into a dead, dry ball of dirt.

Part, if not a large share of it, can be blamed on what I refer to as the human race's handlers or controllers that I will refer to as the underlords. These beings, some not necessarily human themselves, began thousands of years ago with a program of complete manipulation of the newly created human race. First off, thanks to our initial genetic design, it made it necessary for us to have material needs such as constructed shelters to keep warm due to our hairless bodies.

If you have ever watched the movie "The Truman Show" starring Jim Carey (I highly recommend it), the focus was on an individual named Truman who, from birth, lived in a very large, constructed movie set that would serve as his entire world. This was the ultimate reality TV show where, as far as Truman was concerned, the life he was living *was* reality. He had no idea that he was living in an elaborate illusion created to benefit the fictional producers of the "The Truman Show", at Truman's expense! Sorry about this spoiler if you haven't seen it yet but Truman ultimately became wise to his situation, discovering that everything he was ever taught was an illusion and was able to find an exit door and walked out into the real world.

In like manner, we *are* Truman and are living in a completely manufactured reality, all for the purpose of manipulating us for the benefit of the producers who, in this case, are the non-human/part-human underlords. They have succeeded in ways you cannot believe in building this grand illusion. It comes complete with the "good guys" and the "bad guys" to make you think there are only two main points of view. Living in such a "box", we would never realize that there was a third option or way of thinking. As long as we were thoroughly entrenched in this box or artificial environment, we would never have given a single thought that the third option was simply that this "box" was an illusion itself and true reality was just outside of it!

So, how much of our lives is an illusion? For starters, let's talk about material needs and wants. It cannot be argued that humans do indeed have basic material needs to survive: clothing, shelter, weapons, tools to create/build things and cooking implements. I would say, this was what people two centuries ago had as most of their possessions. This was their needs and could make a very small list. Fast forward to today and I would say that the average person's list of "needs" would be many times larger. What may have existed before as a "want" is now considered a need by today's standard of living. Add to that, the creation of new devices and inventions to "make our lives easier" has amplified that "needs list" to the point of ultimately causing massive clutter in our homes. The underlords, by conditioning us to think we cannot get by without the latest and trendiest device, have caused us to become further dependent on their control system. How many times do we hear people say, "I can't live without that!" or "but the Jones' have a new car, why can't we?" It's amazing to see how addictive material possessions can be, especially if you ever watched stores open up early mornings on Black Fridays - the worse of humanity can be witnessed. All this is fueled by the constant drum beat of commercials that pound the message of "Buy! Buy! Buy!" into our heads, turning a "want" into an absolute "need".

Another similar form of conditioning is what we have been taught with regard to being successful in life. We are presented with this ideal of "living the good life!", as if there's a carrot on a stick being dangled in front of us. Shows featuring the lifestyles of the rich and famous bombard us on TV constantly, adding more fuel to our needs list to the point of insanity! Really, how many homes do we really need? Why would we need a huge home for just a few people? Could it be, we have been told from birth that the sign of success is the accumulation of material things or property? Could it be, we have been told from birth that it's important to gather praise from others for our quest of gathering material possessions?

So how do material possessions play into the underlords hands in controlling us? Let's think logically here. Isn't it true that the more we have, the more we have to worry about? Don't material possessions like homes, cars and boats require regular maintenace to

[146]

keep up? What about the paying of more taxes? The more we have, the more it will cost us. And what do we use to pay for things? Money! And if we don't have enough money to pay for our material pleasures, what do we use then? Credit! So what am I driving at here? Well, in order to have the money to cover our credit debts, we need to have a higher paying job.

The key is this: the underlords do not want us to live an independent lifestyle much like many had a century or so ago. By pressuring us to buy and live beyond our means through the use of credit, we would be easy to control by forcing us to work long hours just to survive. Also, by being overly dependent on a high paying job to maintain our lifestyle, we would more likely be submissive to a corrupt system, fearing the loss of that job if we voiced our objections to unjust actions that threatened civil liberties.

Another form of control involving material things is the endless creation of social programs to benefit people by providing them food or money indefinitely, without the expectation of them getting on their feet to become self-sufficient. This too creates a serious dependency on the state for our every need. If one were in this position, he would be hard pressed to refuse the wishes of the state over the threat of having the welfare payments suspended or worse!

Now, let's examine the spiritual side and how the underlords use belief systems to further their control agenda. This is the one area that people will have the most difficulty in dealing with since not all belief systems can be correct, especially if they're supporting opposite points of view. In the politically correct society of today, the issue of religion has become a power-keg and a landmine if one puts down another's faith. So I will approach it from an unbiased view in the form of questions to ask oneself:

Does ones belief system promote a dreadful fear of god and death?

Does ones belief system place demands on being obedient to a god or else suffer the consequences of an unspeakable and eternal afterlife punishment?

Does ones belief system promote the condemnation of others or their actions, who do not subscribe to that particular faith?

If you answer "yes" to any of these questions, you will understand why we have had constant wars since our civilization began. Multiple religions have been purposely created in our artificial environment to fuel our differences, separating ourselves from each other and preventing us from unifying against the real enemy to mankind – the control of the underlords.

Let's go down the rabbit hole a little further and discuss something related to beliefs as it applies to our own human history. In the mainstream, the underlords conveniently gave us a couple of choices in picking a history that best suits individuals. In the religious arena, there's the common creation teaching of Adam and Eve by God and how all of mankind descended from them. Also, we have a version for the atheist called evolution of the species where we, as humans, are the result of a long chain of ever evolving creatures that started in an amino soup of acid in the ocean to our present form. What both of these teachings hold in common is, there is no room for any Extraterrestrial involvement. The underlords go to great lengths to try to disprove their existence by claiming we are alone in the universe. The reason will become apparent as you read on.

Fortunately, a third option is making it's way into the consciousness of people that we are, in fact, created and, we also evolve. Rather, I should say, our DNA is a direct result of creation with the capacity for evolving itself into all the life we see today. As things get slowly revealed, we are learning how today's scientists are already creating hybrid combinations of animals and of plants. It is therefore not a stretch to say that some advanced Extraterrestrial race could be responsible for creating a hybrid hominid that later becomes known as the human race, with the same being true of the Sasquatch. Also, it's getting harder for the underlords to hide all the evidence around the world of past civilizations and technologies, much more advanced than our own, in the building of the pyramids and other structures. Also noteworthy is the finding of giant human-like skeletons from a past era. Accounts of UFOs and Extraterrestrial races and their involvement with our governments can easily be found in the testimony of many key and credible witnesses.

So why are the underlords so intent on hiding the *true* history and origin of the human race? For one thing, the current narrative suits them quite well in their agenda to keep us under their wings. Quite frankly, what they are hiding from us, is all the terrible acts that have been perpetrated on the human race for thousands of years - endless wars created to serve their purposes of financial gain and a way to cull the population down when it gets out of hand. Since the underlords are, by and large, either Extraterrestrials themselves or direct hybrids of such (many being considered the gods of our current religions), they do not want us to know that *they* are the ones responsible for the current mess we're in, especially if the human race wakes up to the deception and massive lies that everyone has been subjected to. They also do not want us to evolve by learning the knowledge that keeps them in power. In a sense, since they only comprise a small number compared to the total population of the earth, they have more to fear from us than the other way around. Can you imagine if, one day everyone woke up and had the knowledge of what was really happening? It would be game over for the underlords and a new day would begin!

The underlords certainly don't want us to have plenty of free time to think about our current situation and get to the point of really waking up to the truth. They want to keep us preoccupied and perpetually distracted with no time to think deeply about anything. Aside from having to work harder for less, due to decreasing wages and inflation, our remaining time is spent more and more in front of what I call the idiot box (TV) to save money from going out.

By watching programs, the underlords found a perfect solution to control us by dumbing us down so that we can't use critical thinking to figure things out. Like an additive drug, people become obsessed with "living within the TV tube", an artificial world that's within the artificial world we're already in! The underlords will go to great lengths to use sports and cop/police/lawyer stories to lock us in.

Under the spell of the "tube", the underlords can now program us to do their bidding, causing us to buy anything they *want* us to buy. Big pharma pushes endless commercials for their newly developed drugs for profit, with safety seemingly a secondary consideration. Lawyers also push endless commercials for promoting lawsuits against the drugs big pharma are pushing that harm people – again for the purpose of profit.

Public opinion can be shaped concerning issues like gun control or mandatory vaccinations. By downplaying news stories concerning how guns have saved people from becoming victims and instead, focusing on stories of how innocent people were killed by guns in the hands of criminals, the underlords can then program the thought that "Guns are bad!" in the minds of the masses. Honestly, the underlords do not want the public to have any means of defending themselves – it makes their lives easier in controlling us!

The same technique is used with regards to vaccines. By failing to report on the stories of how vaccines have either caused deaths or permanently disabled people and emphasizing the ones that claim vaccines are effective and safe, the general public is swayed to accept the taking of vaccines. Again, it's a big plus for big pharma to push vaccines for guaranteed profits.

Speaking of big pharma, another underlord agenda is to get the masses to become dependent on drugs whether legal or otherwise. By impacting your health adversely, we lose much of our personal power and ability to think clearly. This is all for the purpose of preventing us from focusing on the real problem – the control by the underlords!

So, in view of all the points mentioned earlier in this chapter, we have to ask ourselves, "Is this what we call progress for the human race?" People may call this "civilization", however, the Sasquatch would refer to it as "un-civilization". Where we are today was never supposed to happen if things went the way they should have thousands of years ago. We, as humans, would have stayed connected to the earth and all its loving energies. Our lives would have centered on spiritual concerns instead of the accumulation of materials things.

In the Sasquatch's world, the idea of ownership of land and material things is foreign. By being connected to the Earth, we would adapt to the environment through changes of our DNA brought on by those connected energies. So in essence, we never allowed ourselves to continue to evolve as the Sasquatch and all other creatures continue to do.

So what can we do now that the world is in the situation it is? The machinery that runs the corrupt world system is breaking down quickly and the artificial environment that many are living in will dissolve away. Those that hang on to dear life to this world – its egotistical madness, material possessions, a "service to self" instead of "service to others" attitude will fare badly and probably not survive.

This applies to the underlords as well – their control over us exists only because we have allowed it, although unconsciously by most. The more people become aware of their deception, the less control the underlords will have over us!

Those of us who see the writing on the wall need to step back and see that, from birth, our lives were carefully manufactured for us to create an artificial box that many would call their comfort zone. We must first acknowledge to ourselves that the old ways we were accustomed to will no longer work. We need to reinvent ourselves by taking complete personal responsibility for our lives. We need to learn how to be self-sufficient, not depending on a government to take care of us. While learning to do for ourselves, we must keep in mind that the Earth is our real environment and thus, we need to respect her for continuing to provide us with air, food and water. We must not harm the Earth by our ways, we need to take only what is needed for our families or clan/village. Technology is not a bad thing but it must not pollute the air or water, nor should it damage the land. We must remember we share this planet with other forms of life, great and small. All is precious in the sight of God/Source and we must respect everything and everyone.

Those of us who are reading these words and taking them to

heart have to view ourselves as the awakened ones that will bring in a new era for *how* humans will live in the near future and beyond! We need to reconnect to our planet and re-learn the ways we long forgot. Collectively as a race, we need to be humble toward our older Sasquatch brothers for what has been done to our shared home. A time of healing needs to take place as we slowly reintegrate back into a full relationship with our forest friends. This is my sincere and heart-felt hope – I hope it is yours as well...

Chapter 25
Know Yourself - Question Everything!

First understand yourself before you try to make sense of everything else. What does this really mean? Understanding yourself is first acknowledging that we are a product of a life time of mental conditioning (what we've been taught or subjected to). As a result, we subconsciously filter out or reject any information that falls outside of the boundaries of what we have been conditioned to accept (i.e. the existence of the Sasquatch). This becomes the basis behind discrimination between different groups of people. One group has been conditioned from birth to accept a set of ideas presented as "facts" while the opposing group are also conditioned but presented with different or polar opposite ideas that they too accept as fact. Hence the clash results and nothing changes because both groups fail to realize that they have been conditioned to think this way in the first place!

A touchy case in point, especially for religious people, is the differing viewpoints of most religions. How often have we seen the strife and division that results from religious conditioning of different ideas resulting in wars throughout history. It should be noted that religions are the perfect tool for controlling a massive number of people and the elite controllers of our world have made effective use of this. Especially over the last half century, this conditioning is further fragmenting the human race by teaching confusing and conflicting information about everything you can think of.. religion, science, history, philosophy. This has been done, by design, to produce a population that fails to unite on thought and intent about anything important. Today, everyone has an opinion about what is right that varies from everyone else's opinions.

Let's add another form of conditioning that has had the most success - television. Think about this: Why do they call a show on television a "program" or "programming"? Also, think about the kind of programming content that is now popular versus say forty years ago. Can we honestly say that programming today is increasing your mental focus or, is it doing the exact opposite?

The first step is to acknowledge that we all are victims of an aggressive mental conditioning program that has dictated everything we believe, say, or do since the day we were born. We need to realize this before we can do anything else. We must be willing to accept the one fact that everything we have ever been taught may be a lie... everything! Do we think that all the correct facts have been presented to us on a silver platter? How often have we been encouraged to question *what* we believe?

Are we willing to ask ourselves these difficult questions?

If we belong to a religion: How do I know if my religious texts are true? What are the exact origins of the religious texts I use in my faith? When were they translated? What are the earliest manuscripts and who wrote them? How do we know they were authentic?

About government: Does my government work for it's citizens? How do I know if my government is using our taxes to fund secret technologies that could be used against us? Why do we continue to support it financially if it is not working for us? Why don't we replace the entire process if it's not working for the interest of its people? Am I willing to do what is necessary to secure my personal freedom and not be subjected to tyranny?

About science: How do we know if certain authorities such as scientists and doctors are giving us (the average person) correct information? How do we know if there's an agenda to cover up specific information to keep the masses in check for a desired result?

About education: Why is education not given a priority in our community? Why do we have to go into debt to pay for a decent college education? Why does the quality of education continue to decline? How do I know if *what* we are being taught is even correct especially in the fields of science and history?

About business: Why are we living with a monetary system that benefits only a few while the rest live in poverty? Why is there so much emphasis on climbing a corporate ladder just to be downsized in later years? Why has the work environment become so regimented in recent years and your value as an employee has been reduced down to what is called a "human resource"?

Granted, these are but a few questions but I believe this gets the main point across:

Question, Question, Question EVERYTHING!

Personally I started this process a few years ago and I have never looked back! It has been an amazing time of discovery and mental/spiritual evolution for me and it all started with asking the question "Why?"

Is it not time for us to set out on our own journey to start with a clean slate and begin the process of questioning? Do not let fear control us or stop us from discovering the truth. As what was so appropriately stated in the X-Files series, the truth is out there! We just have to be willing to not accept anything we have been taught or heard on face value but diligently do our research to establish for ourselves what truth is.

Chapter 26
The Worldwide Agenda of Fear

Fear! The word alone carries a plethora of meanings to the average person. In many ways the concept of fear literally controls the lives of most people. Allow me to list just a few of them so all can know what I'm talking about.

Immediate fears most people experience:

> Fear of crime committed against us or our families.
> Fear of losing our jobs.
> Fear of not making enough money to meet our expenses.
> Fear of reprisal from an unjust government.

Now add fears that involve the immediate future:

Fear of an economic crash.
Fear of not being able to retire without working.
Fear of war.
Fear of pandemic diseases.

And finally the ultimate fears:

Fear of death whether to our loved ones or ourselves.
Fear of afterlife horrors such as hell
or perhaps fear of no afterlife at all.
Fear of a god who demands fear from its believers.

No one will argue with the fact that we, as a society, have seen a gradual buildup of worldwide situations that have been pressed on the human mind, the result of which has produced a constant state of fear. However you may be wondering why I am referring to fear as being part of a worldwide agenda - an agenda by whom?

Before you can identify the "who", you need to be realize that there is an actual agenda of fear by someone in the first place. The best way to start is to first examine the effects of what fear does to the human psyche which is the totality of the human mind, conscious, and unconscious. So what is fear?

Fear is defined as a distressing emotion brought on by an impending danger whether real or imagined that is perceived to be uncontrollable or unavoidable. However let's first exclude from our discussion what I would call "healthy" fears such as fear of the laws of nature i.e. being too close to the edge of a cliff for fear of falling. Notice in my defining fear there are two points to consider:

1) An impending situation that could be a real or imagined danger
2) Our perception of the situation that causes us to feel that we have no control

Fear is almost always related to our mental contemplation of the immediate or long term future and, depending on how we perceive that future, will determine our emotional response. For us to mentally contemplate something, we have to first allow our minds to gather enough information to get a specific picture of what the situation is and how it can impact us personally. For instance, if a person can be fed enough negative information that ultimately makes him believe there is a horrible situation pending, say a death and doom scenario, then that person will begin to exhibit the emotion of fear. If this emotional state is prolonged over time then other negative emotions will spawn out of fear such as anxiety, dread, terror, fright and panic. As a person undergoes these emotions long term, it will start to close off their ability to see a bigger picture of life and totally paralyze them to the point where they no longer have control of their own life.

Understanding this bit of human psychology, it would not be hard to imagine that, if we really wanted to control a group of people, we would first control what information they would be allowed to see or hear. It would not matter if the information was real or not, as long as it produced a certain perceived reality in the minds of the people. Therefore, if we could create a perceived reality of dreadful impending situations in the minds of the people then they would begin to exhibit fear and before we know it, they become easy to control. They perceive themselves as powerless, even if all of what they think is real is just an illusion. It does not matter if it is just one person or a million that are rendered powerless, an entire population can be controlled this way by just a handful of manipulating controllers!

Can this really be so? Look in the history books and examine what Nazi Germany did to their people through propaganda. The German people were, at first, lulled into an illusion of a Third Reich that would last a thousand years but slowly over a twelve year period succumbed to such fear that no one would dare speak a word against Hitler or his Nazi government. Fear even caused the people to turn a blind eye to what was happening to groups of people such as the Jews in the concentration camps. Ultimately, the fear of death overrode every compassionate human emotion for their fellow man and in the end, the consequences to the German nation were disastrous. Had the German population not succumbed to their fears in the first place and not allowed the German propaganda machine to create a false reality in their minds then Hitler and the Nazi party would have never been allowed to control their population. So it might be said that the German people allowed themselves to be gradually controlled. When it hit a certain momentum of critical mass, the people were unable to stop what happened to them and they had no choice but to see it all the way to the end.

So what about today? Can we see this happening in the world? Can we see this already happening in our country? It would not be necessary for me to point out all the impending situations that the American people are being subjected to now, I have already listed a few earlier in this chapter. But the real questions we need to ask ourselves are...

> Are these situations being fabricated to build a certain perceived reality in our minds and... what kind of picture is being painted from all of this?

> Is it a promising future of security and economic prosperity where all our citizens are treated fairly and are able to provide

for our families without fear of war or crime?

Well, quite the opposite has been painted. No one will dispute that people have become fearful as a result. If that isn't bad enough, more fuel is being added to the fear machine by bombarding us via the news media and movies that capitalize on "end of the world" stories that add to the drama.

Now, let's talk about religion in general for a second. Overall, has it had an empowering effect on the human mind or quite the opposite? Unfortunately, a number of those who profess to be spiritual have convinced themselves that the death and destruction of mankind mentioned in the book of Revelation must become a reality because the Bible said so. Yet, they have become fearful of the very god that promises this destruction, least they fall victims with the rest of the world in a coming Armageddon. They are programmed to fear death for they are told that God punishes bad behavior in a ghastly hellfire for eternity or some other destruction. They are disempowered as individuals when they are told consistently that they are sinful creatures and are powerless to change their immediate situation. All I can tell you, without going into more details, is most religions organizations are merely another form of control based on fear to keep large populations of people in check. This "perceived reality" we have been living under has been going on for thousands of years. Even the history of the human race and the true nature of God/Source has been completed distorted and misrepresented.

Now, let's look at the big picture again and we should start to see that there has been major manipulation of the human race for quite some time. By whom, we ask again? There is enough evidence now to conclude that there is a very small faction of people who have been in charge of the entire world's affairs. Everything has been manufactured or manipulated for the desired result of this elite group. Wars were planned out ahead of time with the winners and losers already predetermined. Visible governments have been set up to give the public the perception that they have power in the voting process when in fact, all this is staged in advance. This is the grand deception and it is easy to live within this manufactured matrix reality and never know it. What we must know is this: fear will keep us trapped within this matrix. Only when we finally step *outside* of it and recognize it for what it truly is, will we know the truth of who *we* really are and who *they* really are.

The first step to self-discovery is to educate ourselves through personal research of what the truth really is and allow our inner intuition to guide us as to whether something is valid or not. Starting a practice of daily mediation will allow that intuition to speak to us from within. As we progress and the true facts become firmly rooted in our being, the fear and our old perceived reality will slowly dissolve away.

Only then will we realize how powerful we truly are as we shape our own reality and not the reality of those who wish to control us!

Chapter 27
Humanity's Time for Decisive Action

As of the past few years, we started down a new period of change that has been accelerating ever since. We are seeing more earth changes such as weather and earthquakes occurring more frequently and the world governments continue to be destabilized. We have plenty of evidence that indicates things cannot continue like this much longer. Most of mankind continues to be totally clueless as to the real meaning of why these things are taking place. Instead of uniting together to solve the puzzle of the big "Why?", everyone continues to divert their individual attention in a thousand different directions. Ever increasing distractions coming from all sources bombard the human race and basically overload the mind with meaningless thoughts that will not change or improve ones life one iota. Confusion reigns supreme as we are dished out different versions of earth's history, conflicting opinions of doctors, scientists, politicians and especially the divisive nature of Earth's religions in finishing off our ability to think logically to make any rational choices about our lives. Finally, after all that, we deal with governments that insist on taking more control over our lives towards the goal of turning us into mindless resources for the state.

So have we had enough? Earlier, Chapter 25 I encouraged all the readers to Question, Question, Question Everything! Here are a few more questions to ponder:

> Are we now ready to make a commitment for change and take the time to truly understand the illusion we have been living under for thousands of years?

> Are we sick and tired of being suppressed and squeezed in from all sides?

> Are we ready to leave all the confusion behind and use the true power of our mind to understand the basic truths of our existence such as...

Who are we really - what is this thing we call consciousness?

What is the true nature of God/Source?

Does religion even represent what science would call the God/Source?

Or is religion an elaborate scheme to control the human race?

Why would we assume that out of the vast billions of planets in the universe that Earth is the only one that has intelligent life?

Why are we waiting for the government to solve our problems?

Why are we waiting and depending on a god to save us?

Why are we ignoring the plight of those in need or dying of starvation in other parts of the Earth and not giving them a second thought if we have plenty?

Why are we watching television when we could better spend time learning something constructive?

Why are we paying for fuel and utilities when alternative energy like solar and wind are available?

Why are we not supporting the effort to create free energy on the planet?

Why are we not questioning everything we've ever been taught?

How do we know that anything we've ever been taught is even true?

They say the average human mind is only utilizing less than twenty percent of what's available...

Why are we not striving to use the other eighty percent?

Why are we giving our power over to others to control us?

It's time for the human race to wake up and contemplate the unthinkable. It's time to educate ourselves now before it's too late.

What do we need to do now as a start in order to get control of our lives?

    Knowledge is everything and the journey to acquire true knowledge is more precious than all the wealth on the planet! It's the knowledge that the world's controllers have desperately tried to keep from us in their quest to keep us in line.

    Do your own research to validate what I have discovered in my own journey. We, as humans, have to start looking within ourselves and listen to our own intuition to really know when something we hear is true or not.

    The future holds promise for humankind, as well as for all life on our planet. The Sasquatch peoples have been instrumental in helping me my entire life to bring me to the point where I was able to write this book.

    It is my sincere hope that the messages written within these pages will benefit each person that reads them and, in turn, helps them to find their *own* personal "journey through the veil!"

Appendix I
The 2012 Event

Below was an article I published on Nov 24th, 2012 on my blog site PleiadianSage.com. In re-reading it while preparing this book, I was amazed how accurate the information was in describing the situation concerning that date and what was expected in the coming years. I am including this as part of the book because the message is especially important to us and is in alignment with what the Sasquatch are revealing to us today:

So What's Really Going to Happen on December 21, 2012?

There has been much said and written concerning this date for over a decade and as a result there is a wide spectrum of information ranging from the "end of the world" to just simply an end to a calendar cycle created by the ancient Mayans. I have personally researched this issue intensely over the past two years and have slowly pieced together elements of truth and separated the rumors and disinformation. However, in order to be totally unbiased, I had to look at every aspect of this subject - the spiritual as well as the science but also other sources that speak in common pertaining to this same time period. Let me start by saying "I don't really know where to start" as this research has taken me on such a journey of mind and spirit that it has totally changed my perception of reality itself. It has led me out of bowels of religious dogma and deep into the science behind the secrets of the universe and brought me in touch with the true spirituality that is within all of us.

First off, this article will just summarize what I have discovered but I will leave the details to those that wish to research further. There is plenty of documented information available whether through hardcopy or internet although I would be cautious of material that merely expresses opinion and does not present much fact. Also, it is important for you to be in touch with your own ability to decide what resonates with you as either fact or fiction. You can, in fact, sharpen this ability of discernment through daily meditation that allows your conscience mind to become more connected to the subconscious.

The first thing everyone needs to know is December 21, 2012 is NOT going to be the end of the world as one would imagine by a nuclear holocaust or meteor collision that results in humanity being toast and the earth becoming a charred rock in space. However, it DOES signify the beginning of major changes on our planet that will ultimately lead to a golden age for mankind. However, much will change, not necessarily right on December 21st, but as we go into 2013, events will accelerate and will become very apparent to everyone that the world we once knew is no more.

There are three categories of change that will take place - physical earth changes, economic/political and the spiritual.

Within the physical, there will be continued earth changes. Even now (and for the past few years), we are seeing the evidence of this change in the strange weather patterns and increased volcanic activity around the planet. We will continue to see tectonic plate movements that will reshape some of our landmasses and coastlines (it's not a bad idea to relocate to a higher elevation when you consider what has happened to New Orleans during Katrina and most recently the East coast with Hurricane Sandy).

Also, there is the ever present danger of an EMP (Electromagnetic Pulse) from the sun that could fry our power grid and some electronic equipment. Scientists know it's not a matter of "if" but rather "when" this will take place. Incidentally, the sun goes through an eleven year cycle of peak solar flare activity and this year [2012] is the peak of that cycle. So just imagine to yourself if electricity was no longer available!

What would that do to today's society as dependent as it is on electronic devices and the internet? Everything would come to a screeching halt and the only thing that would become important is survival: "Where can I get enough food and clean water? How can I keep warm? How will I protect myself from others seeking to take what little I have?" There are thousands in the New York/New Jersey area that are still asking those questions though we do not hear much from the mainstream media covering this on-going story. To get some idea of what life would be like without electricity, there is a new TV series called Revolution on NBC that does a fair job of visualizing this.

Another change that has been building for a while has to do with our economic system and our dependence on a monetary structure clearly out of control. Our lives up until now have been built around the power of currency and the power it has over us. We are rewarded with financial prosperity if we fall in line and are obedient to those that wield even greater financial influence. I'm sure everyone has heard the saying, "He who has the money has the power" - this is not just some cute idiom but indirectly implies a duality of the "haves" and the "have nots" - all based on how much or how little we can have based solely on a monetary system. At this time, this system is like a house of cards that can collapse at any time without much effort. For instance, any of the earth changes I mentioned above can be the tipping point especially if we are hit with a significant Electromagnetic Pulse that will bring down the complete banking system. Once there is an economic collapse, it will cascade in a collapse of the political/governmental systems throughout the earth. We are already seeing the instability of the European nations like Greece and Spain due to the problems associated with the Euro currency. If this becomes a global phenomenon then, from a non-spiritual viewpoint, all becomes hopeless as the world slowly destroys itself through chaos and anarchy.

Lastly and most importantly is the change related to a spiritual shift. The Mayans were clear to point out that the end of this calendar cycle was just the beginning of a new start for humanity - a golden age to the divine feminine. This is a basic change to each and every person's state of consciousness to a new awareness of who we really are and what reality truly is. This will transcend all the current economic/governmental/religious systems and institutions that have divided and polarized mankind for thousands of years. For eons of time, we as humans have been brainwashed into believing that we do not possess the answers to the "why" of our existence. We have been told to look "outside" ourselves for the answers whether it is to a higher god or authority separate from ourselves. Religion has been the worse program of this dis-empowerment by implying we are not powerful beings but sinful creatures that need to be saved by some higher entity. Sin has been translated as meaning "falling short of the mark" and maybe in this sense we, as the human race, fell short by "forgetting" how powerful we really are by allowing a small elite group of control freaks to run our lives! Even now, many of us are waking up to our true nature - the powerful and eternal entities we have always been. We are not simply a human body with a soul - we are in fact a powerful energy light being with the ability to project a physical body and yet exist on so many different planes of existence simultaneously. We do not need anyone to control our reality and tell us that we cannot move forward in the experience we call life. All will become aware of our direct divine connection to the source of all life we call God and realize we do not need any go-between or mediator to make that connection. This is the "shift of consciousness" that will wake up all of mankind to this true reality and, once this takes place, our innate yet dormant powers of creation will activate and we will bring back healing to not only ourselves but also the planet Earth (a living conscious being herself). There will be no more duality as we know it - bad versus good, male versus female, Democrats versus Republicans - all this will be a thing of the past. We will be a new human species - doing everything for the greater good of all, leaving all of ego behind as just a phase of our growth much like when adolescence becomes an adult.

As I said, this is a basic summary of what to expect and it is a fairly accurate depiction of where we are at and what is to take place. Please keep in mind that December 21st, 2012 is just a marker, a beginning of what is to shortly take place. One should use their own senses to see what is happening around them and look from within to know when these changes are taking place. I hope it is the desire of everyone to learn the truth and not to stay in the illusion or comfort zone that some of us desperately attempt to hold on to. Change will happen - it is best to be mentally and spiritually prepared for it!

Namaste,

Samantha Ritchie

Appendix II
Personal Empowerment Message

Everyone starts out in life as blank sheet of paper. Don't let other people write their stories on you but look from within and write your own story.

Become empowered by who *you* really are and not a reflection of someone elses light.

Do not ask permission of anyone for the right to be an empowered individual for you are in fact powerful - we all are very powerful beings!

Take responsibility for your life - become accountable to yourself first before becoming accountable to others!

End the co-dependency on others!

Do not be a follower of other humans for they are your brothers and sisters!

Do not allow other humans to follow you for you are simply their brother or sister!

Stop waiting for the government to solve your problems. Empower yourself first and then work with others who do so likewise and take control of the situation!

Inspire others and stop waiting for someone else to inspire you!

Stop asking forgiveness from your religious authorities whether in heaven or earth for simply being born a human. You were not conceived sinful - you were only told that from the day you were born so that you could be controlled by requiring you to ask for forgiveness and permission to be "you"!

Knock down the walls in your mind and start knowing what true freedom means!

The key to all things possible is within you - it has been there all along!

Unlock the door and change the world as we know it - it is in your power!

Appendix III
Online Resources

Planet Sasquatch YouTube Channel
http://youtube.com/planetsasquatch

Planet Sasquatch Facebook Group
(Search 'Planet Sasquatch' from Facebook)

Planet Sasquatch Blog Website
http://planetsasquatch.com

Barb and Gabby YouTube Channel
http://www.youtube.com/user/Barbshupe1964

Barb and Gabby Facebook Community Page
http://www.facebook.com/BarbNGabby

Samantha Ritchie Contact Information
samantha@txrtech.com

Dr Matthew Johnson Contact Information

10 Crater Lake Ave - Suite 5; Medford, OR 97504

Email: DrMatthewJohnson@yahoo.com

Phone: 541-773-2072

Penosia Sesquai - Audio Analyst

http://www.youtube.com/user/AlaskaGirl1999

Sasquatch Investigations of the Rockies

http://www.sasquatchinvestigations.org

Team Squatchin USA Facebook Group

http://www.facebook.com/groups/TeamSquatchinUSA/

Bigfoot Community Facebook Group

http://www.facebook.com/groups/BigfootCommunity

Thomas D. "Thom" Cantrall - Author

http://www.ghostsofrubyridge.com/

Appendix IV
Young Girl's Report of a Sighting

Back a few months ago, I stopped at a local US Forest Service office to get a wood cutting permit but was told I would need the license plate number of the truck being used to transport the wood which I didn't have. While I was there, I struck up a conversation with the Forest Ranger about my activities as a Bigfoot researcher and the numerous sightings I and others were having in the mountains. Before I left, I told him that I would be back the next day with the needed information to get my permit.

When I returned the following day, the same Ranger was at the front desk and seeing me he said, "There's something I need to show you!" He then handed me what you're seeing on the next page and continued, "What an amazing coincidence! After you left, someone stopped by the same day and wanted to leave us this report from their young daughter about an encounter she had, along with her friends. I hardly ever have anyone report Bigfoot sightings to our office but since you're a researcher, I'll just pass this on to you."

The girl is 10 years old and she dated her report as July 23rd, 2014 [I have blocked out her name for privacy concerns] and the message reads: "Me, Alex and Meara were in the hot tub when I saw Bigfoot! He had medium brown fur. He also had tan skin on his face, his hands, and his feet. He frightened me so much that I screamed and ran." She also provided a drawing showing the Bigfoot as described and being 10 feet tall.

| Bigfoot | 7-23-14 | |

Me, Alex and Meara were in the hot tub when I ⟨saw⟩ bigfoot! He had medium brown fur. He also had tan skin on his face ~~and~~ his hands and his feet. He frightened me so much that I screamed and ran.

Notes

Made in the USA
Coppell, TX
25 September 2020